10-METER FM
FOR THE RADIO AMATEUR

Other Tab books by the author:

No. 859 *The Complete Handbook of Slow-Scan TV*
No. 1120 *OSCAR: The Ham Radio Satellites*

10-METER FM
FOR THE RADIO AMATEUR

BY DAVE INGRAM, K4TWJ

TAB TAB BOOKS Inc.
BLUE RIDGE SUMMIT, PA. 17214

FIRST EDITION

FIRST PRINTING—JULY 1980

Printed in the United States of America

Library of Congress Cataloging in Publication Data

Ingram, Dave.
10-meter FM for the radio amateur.

Includes index.
1. Amateur radio stations. 2. Radio frequency modulation—Amateurs' manuals. I. Title.
TK9956.1516 621.3841'52 80-14490
ISBN 0-8306-9933-3
ISBN 0-8306-1189-4 (pbk.)

Cover photo of Comtronix—FM80 courtesy of KonaCom, California.

Preface

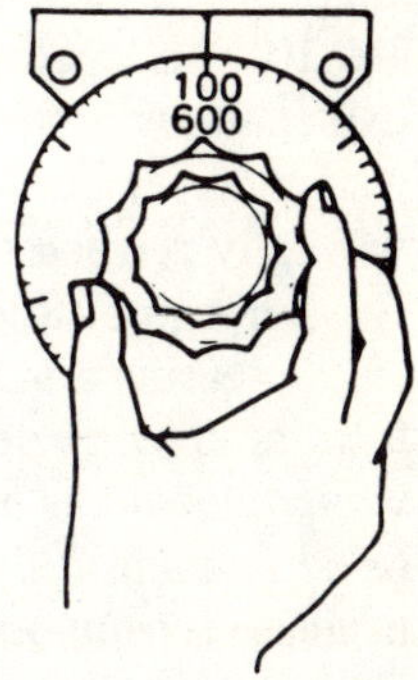

A radio amateur casually relaxes in his equipment room during the early evening hours of late Summer. As he thumbs through a recent ham radio magazine, the squelch breaks on a nearby transceiver. The calling station identifies himself and gives his location as New Zealand. Elsewhere across the United States, a radio amateur sporting a small 2-meter hand-held transceiver walks down a city street while exchanging greetings with another amateur in Japan. On another occasion, an amateur in Europe is heard contacting amateurs in Florida and the Carribean via a U.S. based repeater. The channelized operations of these FM activities provide a flexibility which allows their users to enjoy long distance communications in a reliable and professional manner. Fantasyland? Indeed not. These experiences, and many similar ones, are happening each day in the exciting new frontier of 10-Meter FM.

Situated at the upper end of amateur radio's high frequency allocations, 10 FM supports a vast array of worldwide activity which resembles a super 2 meter FM band with unlimited capabilities. "Direct" communications on "simplex" frequencies are complemented by numerous remote base and repeater setups, all of which provide outstanding performance in this FM wonderworld.

Contrary to popular belief, FM operations are not restricted to local communications. The frequencies usually employed for FM are above 100 MHz, and this range confines the communications distance. Narrow Band FM, which occupies no greater bandwidth than an amplitude modulated signal, can be transmitted on 29 MHz, and it may be propagated thousands of miles.

This book describes the exciting world of 10 FM in a straightforward and easy to understand manner. Since all radio amateurs possessing a General Class or higher License may operate 10 FM, special efforts have been taken to center technical explanations around this area. If your amateur radio interest has waned during recent times, it's highly possible that 10-Meter FM will fully refresh that original interest, and return your personal Golden Age of Radio to reality.

Creating a state of the art book on any new communications frontier is a very delicate matter. Progress continues at a rapid rate while the book is being published. While all information presented herein should reflect the present status of 10 FM, the reader is urged to consult present monthly amateur magazines for any recent changes affecting this new frontier of amateur radio.

I would like to thank the following individuals for their support and assistance with information presented in this book: David O. Findley, N6DF, and John E. Portune, WB6ZCT, for the description of the WR6BDG repeater, and Dr. Don C. Miller, W9NTP, for information on operating parameters of Medium-Scan TV.

Thanks, also, to the following manufacturers for information and assistance provided for this book: Tom Kikuchi, General Manager, Comtronix, Inc.; Glen Whitehouse, Cushcraft Antennas; Martin Jue, MFJ Enterprises; Communication Specialists; DSI Company; Larsen Antennas; Newtronics Corporation; Yaesu Radio. Finally, my most sincere appreciation to my XYL, Sandy, WB4OEE, for assistance typing this manuscript.

Dave Ingram, K4TWJ

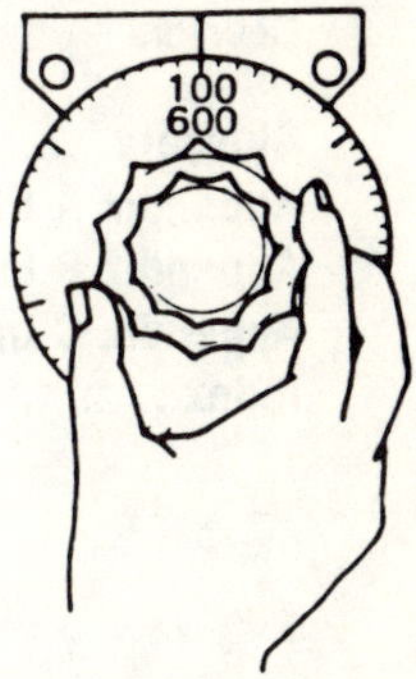

Contents

Chapter 1 Introduction to 10-Meter FM

Until recent times, the majority of amateur operations centered around a specific operating area, communications desk, or an extensive mobile setup. The amateur radio operator was confined to this locale when pursuing various activities like DXing, message relaying, or merely relaxing while enjoying periods of casual conversation. This proper practice was, and is still, characterized by the many radio amateurs "disappearing into that certain room" for a couple of hours of operating enjoyment during any available opportunity.

Occasionally, a mutual feeling of being removed from usual family activities motivated some amateurs to move one or two vital pieces of equipment into the living area or den. Usually, the station's high-power amplifier and many accessories were left behind during this "in house" move. Soon thereafter, this concept proved to be less than satisfactory. Some amateurs handled the resultant situation by constructing furniture-like desks to also support amplifiers, SSTV gear, antenna tuners, etc. Others built elaborate, enclosed cabinets with flip-down desk sections and equipment compartments. Obviously, the enjoyment of lounge-chair operations rivaled the less comfortable in-shack activities. Mobile operations were also entangling during these times, since massive transceivers occupying a large part of the vehicle's prime space were necessary. The time required to tune these rigs was another major distraction to many mobile operators. Several

Fig. 1-1. Dave Greenway, VE2WI, of Quebec, Canada, is an avid pioneer of both 10- and 2-meter FM activity. His present equipment consists of the Yaesu FT-227R and the FT-901DM. Dave is presently pursuing authorization to operate a 2-to-10 meter repeater in his country (photo by VE2FBT).

amateur equipment manufacturers recognized these obstacles and developed compact, minimum-tune hf (high-frequency) transceivers for fixed, portable and/or mobile operations. This approach proved a blessing for many wealthy amateurs, but it left others seeking less expensive alternatives. See Fig. 1-1.

Then came the advent of 2 meter FM, and amateur activity flourished. Mountaintop located repeaters, built and supported by a group of local 2-meter FM operators, gave all users the benefit of an elaborate setup on a shared basis. A small, efficient 2-meter FM unit could easily be slipped beneath the dash of any automobile, or placed beside a family room lounger for serious or casual operations, anytime one desired. The squelch-muted 2-meter radios could be enjoyed for extended periods without disrupting one's usual activities. The "switch on and operate immediately" craze quickly gained popularity, and the amateur 2-meter band skyrocketed to worldwide fame, (Fig. 1-2).

During this same period, 10-meter FM operations were being pioneered and enjoyed by a miniscule group of amateurs throughout the United States and Europe. The desires of this group were (and still are) quite similar to that of all amateurs; they wanted the operating convenience of FM, the mobility of small rigs, and the capability of group-shared repeaters. One other exclusive asset of 10 FM which attracts a rapidly increasing number of radio amateurs is its DX capability. The ability to switch on a compact 2-meter style rig and, without any tuning, contact others thousands of miles away is truly unsurpassed. A typical 10 FM mobile setup is shown in Fig. 1-3.

After thus serving a number of years in pioneering and development stages, 10 FM is now gaining a popularity which promises to rival or surpass 2-meter FM in activity. Each day, amateurs in North America, South America, Japan, England, etc. exchange greetings on 10 FM in the same casual, yet reliable, manner that's enjoyed on 2-meter FM. Although not yet as commonplace as 2-meter FM counterparts, 10-meter FM repeaters and remote-base setups really prove their worth. These vantage-point machines provide group users with 300-to 500-mile coverage during "dead band" periods, and yet they also provide fantastic DX capabilities when the 10-meter band is "open." While many 2-meter FM repeater groups deplore DXing activities, the inverse is true of 10 FMers. Likewise, the prime worldwide 10-meter "direct" frequency of 29,600 kHz is diligently monitored by FM DXers using base stations, mobile rigs, remote bases, and 2-to-

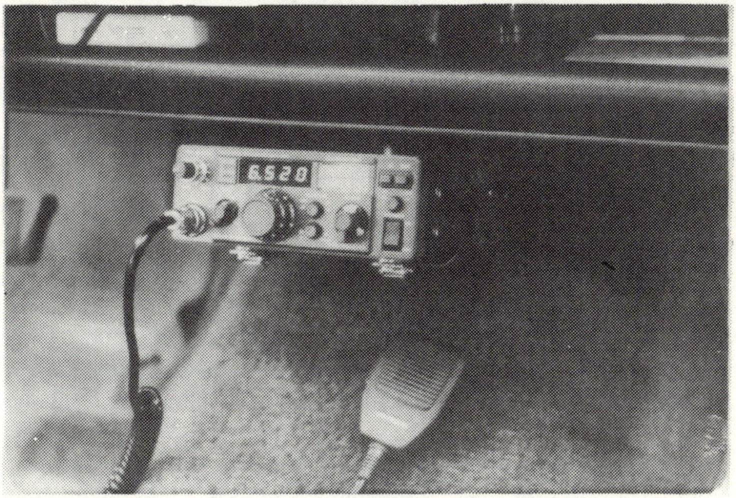

Fig. 1-2. Compact 2-meter FM transceivers similar to the units shown here created mass interest among today's on-the-move radio Amateurs. Their channelized operation and pretuned circuits allow mobile activity to be fully enjoyed. Unfortunately, the range of these VHF units is usually less than 100 miles.

Fig. 1-3. Contrasting with the 2-meter unit shown in Fig. 1-2, this Comtronix FM-80 10-meter transceiver is capable of worldwide communications on a large number of channels. Transmitter offset for 10-meter repeater operation is provided by a push-pull switch on the squelch control. This Comtronix unit represents an inexpensive, yet fully effective means of enjoying the fun of 10-meter FM.

10-meter repeaters. Any number of amateurs can enjoy this activity, either through their individual units, or via joint-sponsored "super machines." The unique experience of working distant countries through a 2-meter hand-held unit is yet another dream made possible via 2-to-10-meter remote-base setups. Thus, considering all aspects, we can truthfully say that 10 FM is a wonder world come true, and this world is open to all amateurs holding a General Class or higher FCC license, (Fig. 1-4).

ASSETS UNLIMITED

The natural flexibility of 10 FM allows this mode to be enjoyed in several unique ways. The most logical means of summarizing these operations equates 10 FM with the DX pleasures of 20, 15, or 10, meters, and the operational conveniences of 2-meter FM (Fig. 1-5).

Fixed-location setups are not confined to sitting by a rig and tuning the band for DX. One merely turns his transceiver on an established frequency, checks the squelch setting, positions his antenna, then goes about other business until the activity starts. Many amateurs, finding this concept extremely convenient, allow their squelched rigs to monitor specific frequencies, like 29,600 kHz, continuously. As a result, we have an "international network" on 10 FM which stands ready to do its duty whenever the occasion arises. Since 10 FM equipment doesn't require a watchful eye for maximum enjoyment, amateurs also find an added convenience in dividing their in-shack time between other modes and 10 FM. While the situation hasn't yet appeared where a station operator accidentally grabs the wrong mike and calls a 20-meter SSB DX station via 10 FM, this possibility does seem to exist. See Fig. 1-6.

The fixed station's linear amplifier may be used for 10 FM operations if desired, however the operator should be aware of FM's demanding duty cycle before operating at high power levels. A continuous carrier is produced during the full time a station transmits FM. This condition is more demanding than regular CW operations because final amplifier tubes or transistors can't cool as they do between dots and dashes. Consequently amateurs should limit their amplifier's power (by controlling input drive signal) to approximately one-fourth its CW rating to assure trouble-free operations. An input of 250 watts, however, is definitely enough power for any and all 10 FM operations.

Fig. 1-4. Numerous foreign radio Amateurs appear on 10-meter FM every day. Anders Bjorkman, SM5AA0, of Sweden is a staunch supporter of this mode. He truly enjoys contacting other 10-FMers. Like many foreign Amateurs, Anders' rig is the popular Yaesu FT-901DM all-mode tranceiver.

Fig. 1-5. One unique asset of 10-FM is reflected in the use of conventional tri-band beam antennas for communicating with specific areas. Vertical antennas are often used when monitoring for band openings. Here we see KL7IYX (top of tower) installing a Cushcraft ATB-34 antenna, with KL7IYH and KL7IZV assisting. All three Alaskan Amateurs enjoy operating 10 FM as well as 20, 15, and 10-meter SSB.

Several options are also possible with remote-base setups. The 10-FM equipment, including the station's high-power amplifier and triband beam, may be operated directly, or (current FCC regulations withstanding) a 2-meter or 70-cm FM unit may be interconnected with this setup for remote-base functions. The range of this remote setup can be restricted by the 2 meter antenna size/location and rf output power. Such arrangments provide the 10 FMer with neighborhood or in-house mobility while also using the main-station gear for cross-country contacts or DXing. An increasing number of U.S. amateurs are presently using these techniques while operating the international direct frequency of 29,600 kHz, and the results have been extremely rewarding.

MOBILE USE

Mobile operations on 10 FM are a refreshing change from aspects like 20 meter SSB or 2 meter FM activities. Fixed-tuned

10 FM equipment can be activated by the auto's ignition switch, yet one seldom finds a conglomeration of station signals blasting forth and creating instantaneous havoc. While some mobiling amateurs enjoy this "relaxed DXing," others add 10-meter scanner capabilities, or include 2-meter units with their array of under-dash gear. One extremely beneficial aspect of using both 2 FM and 10 FM mobile setups is (again, current FCC regulations withstanding) interconnecting the two units to form a remote-mobile setup. An integrated package consisting of a 10 FM unit, 2-meter FM unit, and a "patch unit" for separate/combined operations, for example, can produce some interesting possibilities. The mobiling amateur can operate two bands while in the vehicle, plus he can let his 2-meter hand-held serve double duty and work through the mobile 10-meter setup while portable. Ample precautions should be taken, however, to assure that a discrete 2 meter frequency and a limited range of remote-access is utilized.

CBs MODIFIED

Amateurs desiring the ultimate in custom mobile installations might consider investing a few hours time modifying an in-dash

Fig. 1-6. SSB and 10-meter FM activities blend smoothly in this efficient setup belonging to Steve, KL7IYX, in Anchorage, Alaska. The Yaesu FT-901DM is the main rig, and the antenna tuner adjacent to it is used for lowering the SWR at 29 MHz. When Steve isn't sitting at the desk, the rig stays squelched and monitoring an FM channel such as 29,600 or 29,620 kHz.

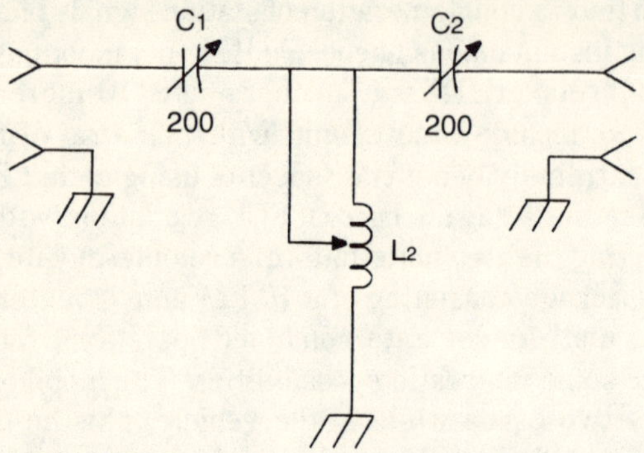

L1: 23 TURNS #18 ENAMEL WIRE WOUND ON PILL BOTTLE OR FORM WHICH IS 1½" DIAMETER AND 2" LONG.

TAP FOR ALLIGATOR CLIP OR ROTARY SWITCH AS FOLLOWS:

TOP OF COIL
#2 1 TURN FROM TOP
#3 2 TURNS FROM TOP
#4 3 TURNS FROM TOP
#5 7 TURNS FROM TOP
#6 11 TURNS FROM TOP
#7 13 TURNS FROM TOP
#8 15 TURNS FROM TOP

Fig. 1-7. Schematic diagram for the antenna tuner described in the text. L1 is 23 turns of No. 18 enameled wire wound on a pill bottle or similar form which is 1½-inch diameter and 2-inches long. Tap points on coil are as follows: 1—top of coil; 2—1 turn from top; 3—2 turns; 4—3 turns; 5—7 turns from top; 6—11 turns; 7—13 turns; 8—15 turns.

AM/FM/8 Track/CB unit to include 10 FM capability. Many of these in-dash units provide "CB priority" which allows the operator to enjoy music of his choice until a CB (now 10 FM) station breaks the squelch. In order for these units to be used on 10 FM, three primary changes must be performed on the "CB section." First, the CB's i-f output must be changed to drive an FM detector rather than the AM detector. Next, the transmitter's modulation system must be changed to employ a simple varactor frequency-deviation system, and the AM capabilities bypassed. Finally, one or two of the CB crystals must be swapped to permit operation on the high end of 10-meters. Modifying multi-mode in-dash equipment is similar to working on compact hand-held gear and its "layered construction" requires a reasonable amount of patience. The final results, however, are definitely worth the effort. A 30- or 40-watt class-C rf amplifier would round out this installation.

If a group of amateurs in a specific area are interested in trying 10 FM on a shared basis, a single group-acquired 10 FM transceiver can be interconnected with a 2-meter or 70-cm transceiver for each person to enjoy. A tone control system should be incorporated with this remote base, thus allowing strict control of the 10 meter link. Likewise, ample precautions should be taken to assure that local conversations are not retransmitted on 10 meters.

Since 10 FM frequencies are situated at the "high end" of 10-meters, a simple antenna tuner similar to the one illustrated in Fig. 1-7 is often used to reduce SWR to a very low level. Comparable tuners are distributed by several amateur equipment manufacturers. As previously mentioned, the duty cycle of FM shoudl be respected when using antenna tuners. A 1 kilowatt SSB antenna tuner shouldn't be used at power levels exceeding 250 watts, or the tuner's coil and the antenna's balun (if used) may be fried to a crisp. Should you elect to "push" the power levels, remember that rf heating effects are cumulative. Short transmissions will thus be inevitable.

All aspects considered, 10 FM is one of amateur radio's most exciting frontiers of communications. The ratio of operating time

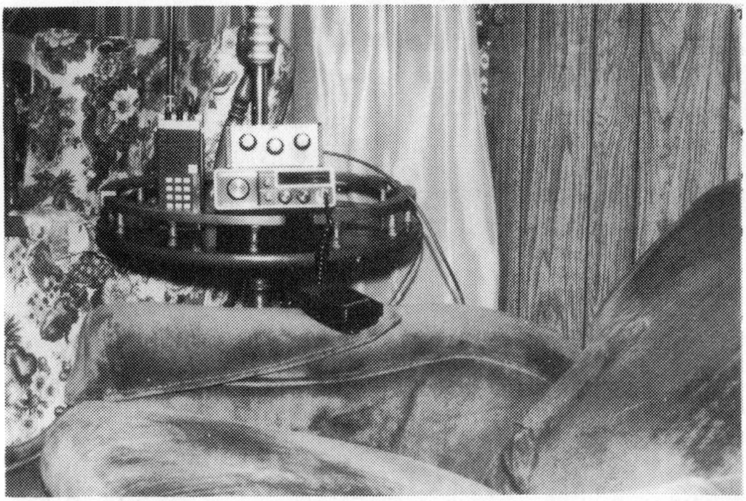

Fig. 1-8. A cozy portable setup for weekend enjoyment is this 10-FM arrangement placed by a lounger in a den or family room. A 2-meter hand-held unit adds a little "flavor" when 10 meters closes for the evening.

to meaningful contacts is very high, and this mode can be enjoyed when usual operating times are unavailable (Fig. 1-8).

LISTENING ON 29.6

One of the most logical means for unsuspecting amateurs to experience the excitement which happens every day on 10 FM is by listening to one or more of the popular 10-meter FM channels. Recently, for example, SM5AAO in Sweden was heard communicating with SM5ANY through the Las Vegas, Nevada repeater. Later that same day, KL7IYX in Alaska communicated with JA7OWB in Japan through a Colorado based repeater. Early during many Saturday mornings, K4GTQ in Alabama has joined a large "ham breakfast" group in Oregon via their 2-to-10 meter linked repeater. Amateurs visiting Las Vegas are often surprised to find their 2-meter hand-helds have world-wide range. However, an overseeing local amateur quickly explains that their unique machine is interconnected to a vantage-located 10-meter FM setup. The number of active remote base systems operating on a 10 FM channel at any one time is readily apparent when a DX station suddenly appears on that frequency. Tones of various durations and frequencies often "pile up" as these setups are energized and activities begin. The one awe-inspiring aspect, however, which has converted many staunch 2-meter FMers to 10 FM, happens when a squelched rig suddenly comes to life with an out-of-country repeater identification. If the 10 FM unit has been quietly sitting for a few minutes time, this sudden blurp sounds exactly like 2 meter FM, but the impact of its DX capability is hard hitting and long lasting.

MEDIUM SCAN TV—A FRONTIER EXPANSION

It may be difficult to visualize a frontier within a frontier but that's the story of medium-scan TV. This unique amateur frontier initially evolved during 1978 when several slow-scan television operators began thinking of a more efficient video communications method. Although slow-scan TV provided a means of exchanging pictures on a worldwide basis, its necessary bandwidth restrictions eliminated the portrayal of motion during each picture. This trade-off was well justified, since previous television transmissions required an extremely large bandwidth, which restricted their use to Ultra High Frequencies. Amateur television enthusiasts, however, wanted to pursue SSTV development further, and acquire the capability of motion. This dream seemed like an impossible feat, since the portrayal of motion expanded bandwidth requirements far beyond allowable tolerances for frequencies below 30 MHz. Obviously, a number of acceptable compromises and a combination of modern technological breakthroughs would need to be included in such a system. Then ideas began formulating into workable concepts, and a system took shape (Figs. 1-9, 1-10, and 1-11).

Since amateur TV communications seldom require large quantities of information change (motion) between successive pictures, a full-motion system was not mandatory. Assuming, for

Fig. 1-9. This memorable SSTV picture shows Io, one of Jupiter's moons passing in front of its mother planet during the March, 1979 flyby of the Voyager I space probe. Io is the center, with the bright clouds of Jupiter in the background. Many Jupiter pictures were relayed to slow-scan TV operators around the world by the Jet Propulsion Laboratory Amateur Radio Club Station, W6VIO. As an outgrowth of this and similar SSTV activities, several Amateurs are now pioneering Medium-Scan TV in the 10-FM subband of 29,100 to 19,300 kHz. This new mode will allow portrayal of motion in nearly real-time, rather than the frozen frame picture of SSTV.

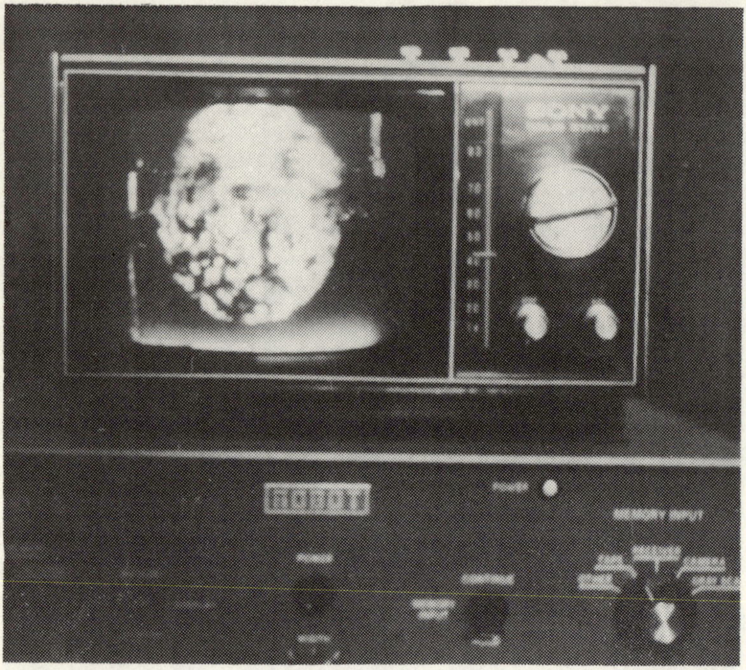

Fig. 1-10. Slow-Scan Television may become an integral part of an Amateur setup with the addition of a digital scan-converter and a display unit such as a closed-circuit TV monitor or a conventional TV set. A simple, one-transistor oscillator allows the scan converter's output to be tuned in on the Sony TV set in the same manner as regular TV stations. Thus, no modification to the set is necessary.

example, the operator facing the tv camera is talking, his mouth occupies less than 10 per cent of the composite scene. Thus it was concluded that medium-scan TV would be limited to approximately 10 per cent motion between *successive* pictures.

Horizontal and Vertical scanning rates were next revised to permit limited motion while reducing RF bandwidth. Since these scanning rates could not be reduced sufficiently to accomplish narrow-band motion TV, a double-interlaced system was instigated. Conventional (fast-scan) television employs vertical interlacing only. This means that two complete vertical scans are required to produce a complete picture. In a double-interlaced system, two complete vertical scans and two complete horizontal scans are required to produce a complete picture.

The medium-scan TV results produced thus far allowed transmission of limited-motion pictures at a rate of 3.75 fields per second. The bandwidth required for these transmissions was 35 kHz. Next, Special Temporary Authorization (STA) from the Federal Communications Commission, to permit transmission of these 35-kHz signals in the 29,100 to 29,300 kHz region of 10 meters, was secured by the pioneering medium-scan TV amateurs. This group consisted of Mike Tallent, W6MXV, Bob Stone, W3EFG, George Steber, WB9LVI, and the original designer/instigator, Dr. Don Miller, W9NTP. Since Dr. Miller had been involved in similar research projects for government applications, his knowledge was directly applicable to medium-scan TV developments. Dr. Miller's close association with the American Radio Relay League and with many overseas amateur television enthusiasts was another prime factor in his position of spearheading the medium-scan TV project.

Research began by checking the RF channel for communications compatibility, and expanded into application of modified slow-scan TV gear for use in medium-scan TV. As this information is being written, the first fully operational medium-scan TV systems are being revealed by their designers. Additional MSTV pioneering and developing steps are anticipated during the forthcoming years. It is hoped that during this time, many amateurs will

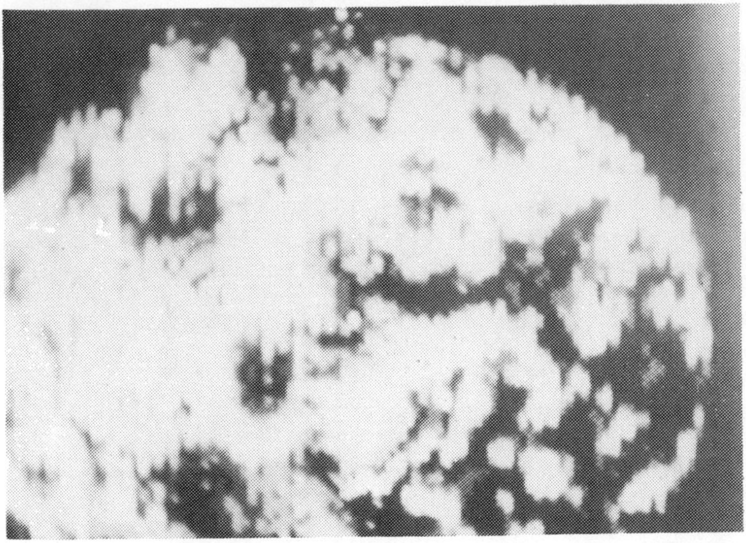

Fig. 1-11. This SSTV picture of Jupiter's moon, Ganymede, was received by the Jet Propulsion Laboratory in a transmission from Voyager I, and retransmitted on the 20-meter band by their Amateur Radio club station, W6VIO. Voyager was approximately 4 million miles from the moon when this photograph was taken. Slow-scan television truly expands Amateur's horizon.

be able to assemble inexpensive P-7 cathode-ray-tube monitor units, and share the excitement of this unusual mode of visual communications.

Medium-scan television activity is centered around 29,150 kHz, and a MSTV beacon transmitter operated by W9NTP may be operational on that frequency as you read this book. Exact, up-to-date details concerning medium-scan TV may be secured via the International Slow-Scan TV Network, which meets each Saturday at 1800 GMT on 14,230 kHz, or via mail from Dr. Don Miller, W9NTP, Box 95, Waldron, Indiana 46182. A large, self addressed-stamped envelope is required for such information.

Chapter 2
The Assets of 10 Meters

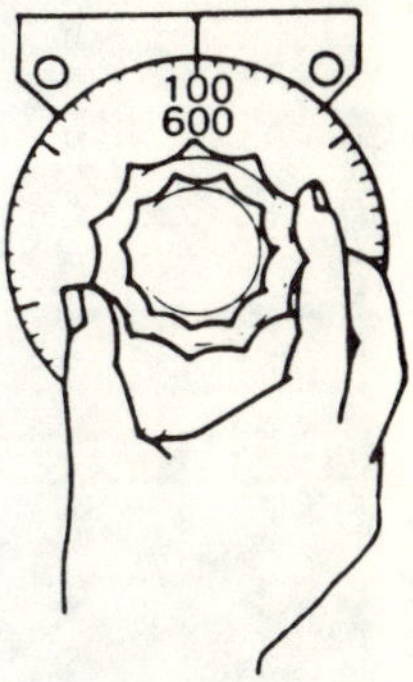

Situated mid-way between High Frequency and Very High Frequency allocations, the 10-meter band offers a wide variety of features for all radio amateurs holding General Class or higher grade licenses. Various segments of this band provide space for activities on CW, SSB, Amateur satellite operation, slow-scan television, medium-scan television experimentation, and narrow-band frequency modulation. Additionally, there are a number of low-power beacon transmitters operating at various points in the 10 meter band, their purpose being to alert Amateurs of band openings into various geographical areas. There's ample "leg room" for all modes of operation on 10-meters, and operators using this band on a regular basis find its "directional effects" highly beneficial in reducing interference.

Although a little-known fact, 10-meters was also used for moonbounce experiments several years ago. Realizing that moon-reflected signals experience approximately 260 dB overall attenuation, one quickly realizes the advantages of low noise, short wavelength bands like 10-meters.

Massive antennas and large amplifier tank circuits are not necessary for 10-meter operations, and RF power requirements for 10-meters are also minimal (Fig. 2-1). Since DX band openings usually center around late mornings and late afternoons, operating 10 FM doesn't necessitate deleting other Amateur pleasures like evening DXing, etc. Traveling amateurs, moving between cities

Fig. 2-1. High power levels are neither necessary or common on 10 FM. The majority of 10 FM operators use less than 100 watts. The 18 dB over S-9 signal being received on this Yaesu tranceiver in Alabama emanated from SM5AA0 in Sweden.

during daytime business hours, find 10 FM particularly enjoyable for brief QSOs between client calls and appointments.

Ten-meter FM operations also blend quite smoothly with occasional activities like hiking or camping. During such jaunts to desolate areas, the enthusiastic Radio Amateur often finds himself out of 2-meter repeater range. Naturally, large units for "low band" operations would be difficult to backpack and power during those times. A small 10-meter FM unit and comparably sized 12-volt motorcycle battery, however, would permit the camper to enjoy long distance communications along with his fishing, or during the peaceful times around dusk. A longwire antenna, thrown over available tree branches, and adjusted with an antenna tuner, would complete this brief setup. Unlike lower frequency Amateur bands, 10-meters is especially adaptable to antenna tuner/random length antennas,(Fig. 2-2).

PROPAGATION OVERVIEW

The capability of long distance communications with low power has been a major attraction of 10 meters for many years. Quite often, exotic DX stations seek these "green pastures" to avoid the high power and interference levels experienced on the Amateur 20 and 15-meter bands. Consequently, the relaxed atmosphere of 10 FM supports a common feeling of international friendship and goodwill. Operators have a chance to explore each other's interests, lands, etc. rather than merely exchanging signal reports and formal greetings.

During the mid-1950s, 10-meters was a beehive of activity as Amateurs used low-slung dipoles or window antennas, with low power CW and AM equipment, to establish DX records. Mobile setups consisting of a relatively insensitive broadcast-band converter and 3 or 4 tube dynamotor-powered transmitter were considered state of the art. Many Amateurs using such equipment reliably contacted areas like Hawaii, Alaska and Japan during early evening hours.

During the mid to late 1960s (the following sunspot cycle),10-meters again created tremendous excitement throughout the world. Amateur stations using Viking Rangers, Heathkit DX40's

Fig. 2-2. This knapsack array of Amateur FM equipment recently afforded outstanding pleasure during a weekend jaunt to the Smoky Mountains. Items shown are Cushcraft 10-meter Ringo antenna, MFJ longwire tuner, SWR bridge, Comtronix FM-80 10 FM trnasceiver, Yaesu 2-meter FM portable, pack full of antenna wires, and a 12-volt motorcycle battery.

(70 watt transmitters), and "plumber's delight" rotary dipoles or "ZL special" beams, enjoyed solid communications with areas ranging from Saudi Arabia and Antarctica to Korea and the Philippines. Numerous times, this band outranked 20 or 15-meters in DX popularity and use.

During late 1978 (present sunspot cycle), 10 meters again began "showing its colors". As this information is being written, the DX capabilities of 10-meters are fantastic, and propagation forecasts are predicting even better results during the next few years! High RF power levels are still unnecessary for 10 meter communications, and over sixty percent of the 10 meter operators use "barefoot" transceivers. During mid-mornings, Amateurs throughout the United States reliably communicate with European and South American areas, with prime "skip conditions" spreading westward like an imaginary blanket; its hypothetical edge (and optimum DX time) occurring approximately three hours after one's local sunrise. Approximately four to five hours later, 10-meter propagation merges to East-West paths and permits Africa/South-Pacific to U.S. type of communications. As late afternoon and early evening hours cover the United States, communications with Japan and South Pacific areas occur, with exceptional signals at both ends of this broad path. Approximately three hours after sunset, depending on monthly propagation variables, 10 meters bids adieu and "folds" until the next morning.

Numerous Amateurs operating transceivers in the 20-watt category are enjoying mobile DXing on 10 meters - and low power 10 FM'ers, operating the 29-MHz range, are experiencing similar pleasures. Anyone questioning the credibility of the previous statement need only check the large number of Japanese stations operating on 10-meters daily, since their most popular Amateur license class limits operators to low power levels.

EQUIPMENT RESOURCES

Several options are open to Amateurs interested in operating 10 FM. We will briefly consider some of the possibilities at this time; and in-depth discussion of each method is presented elsewhere in this book.

One of the most elaborate and efficient amateur transceivers presently featuring FM operation as "standard equipment" is Yaesu's FT901DM. This HF transceiver performs magnificently on 10 FM; it may be operated fixed, mobile, or portable, and its 80-watt input power level (FM mode) is quite sufficient for

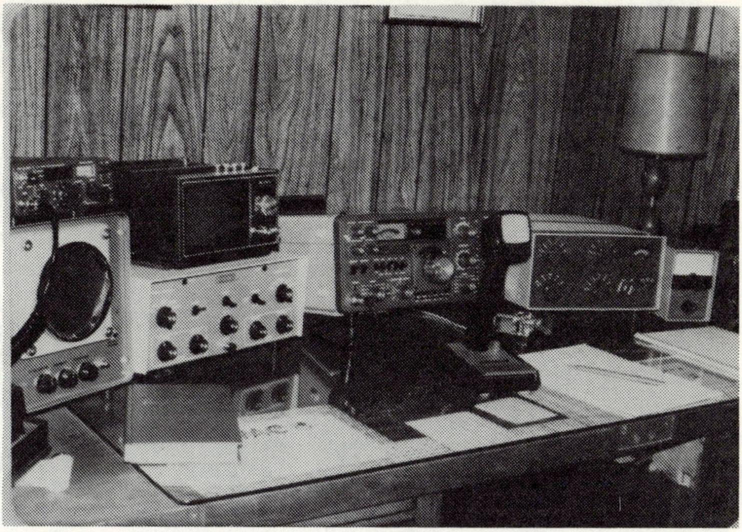

Fig. 2-3. Several modes of Amateur activity are included in the author's main operating position (4TWJ). 10 FM is provided by the Yaesu FT-901 DM transceiver and optional rf amplifier. Other activities provided for are slow, medium, and fast scan TV, OSCAR satellite communication, 2-meter FM, and DX chasing. Additional equipment in the room wouldn't fit into this photograph.

worldwide communications. Any 10 FM "direct frequency" or "repeater split" can be instantly programmed into the FT-901DM via front panel pushbuttons. See Fig. 2-3.

Some older Amateur equipment, modified for FM operation on 10-meters, may be purchased at Amateur auctions or hamfests. If their FM adapters are missing or irrepairable, simple homebrew replacement units can be constructed. Usually, a simple FM discriminator or ratio detector will allow the receiver to function on FM, while a varactor-modulator can be added to the transmitter (Fig. 2-4).

Many modern Amateur transceivers may also be modified to operate 10 FM using the previously described technique. Several Amateurs in the United States and Europe, for example, have modified Yaesu FT101 series, and Kenwood TS520/820 series, transceivers for operation on 10 FM.

As of early 1979, the majority of Amateurs operating 10 FM use modified commercial equipment obtained from surplus outlets (Fig. 2-5). These units include Motorola and GE units of both tube and transistor design. The use of converted business radios for Amateur 10 FM activity closely resembled the initial phase of

Fig. 2-4. A staunch performer on 10 meters, Collins receivers similar to this 75A-2 may often be secured quite inexpensively at Amateur conventions and hamfest flea markets.

2-meter FM activity. Assuming 10 FM follows a similar growth pattern, activity will soon flourish. Surplus commerical units will become widely distributed by large mail-order distributors, and numerous Amateur manufacturers will begin producing multi-channel 10 FM transceivers (Fig. 2-6).

Another source of equipment which should not be overlooked by enterprising 10 FM'ers is Citizen-Band transceivers. Numerous articles on simple 10-meter conversions for CB sets have been published in the monthly Amateur magazines. Basically, this procedure consists of changing hetrodyne-oscillator crystals, then realigning the transmitter and receiver. The only additional steps necessary to place these CB units on 10 FM involves changing the receiver's AM detector to an FM detector, and substituting a varactor-modulator for the AM modulator.

ANTENNA FACTORS

There are an unlimited number of efficient radiators which may be used on 10 FM. The fixed station's triband beam, in conjunction with either a homebrew or commerical antenna tuner, usually produces excellent results at these frequencies. Commercially manufactured verticals, complimented with a few short radials, are also outstanding performers on 10 FM. Non-directional

Fig. 2-5. Commercial "garage sales" of business radios and electronic equipment are often stocked with FM units which can be modified to 10 FM. Occasionally, this equipment is priced by the pound—truly a buyer's delight!

Fig. 2-6. This converted Motorola radio operates on 10 FM in fine style. Total cost, including a deluxe microphone, was less than $50.

Fig. 2-7. This Cushcraft 10-meter FM Ringo is an outstanding antenna for either fixed station or portable use. Its omnidirectional coverage assures being abreast of all 10-FM action, and its gain permits reliable Amateur communications. Ringo shown here is mounted on TV mast, while a triband beam is approximately 30 feet away on a crank-up tower.

antennas like ground planes or verticals are particularly beneficial when monitoring 10 FM, since narrow beam-width, directional arrays may prevent signals that arrive from the side from breaking the receiver squelch. See Fig. 2-7.

Longwires, rotary dipoles, and other unity-gain antennas also perform well on 10 FM. Since interference problems are minimal on 10 FM, longwires and rotary dipoles are popular antennas.

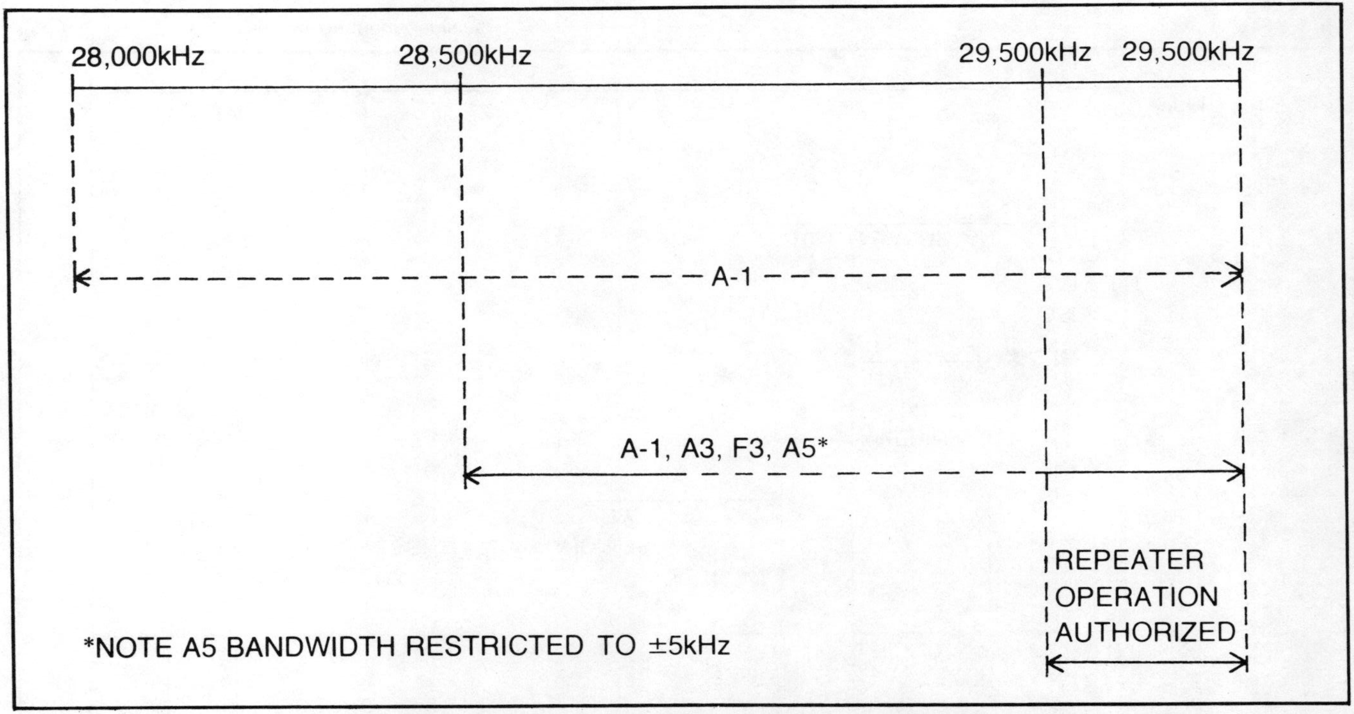

Fig. 2-8. U.S. Amateur frequency and mode allocations authorized by the Federal Communications Commission (FCC).

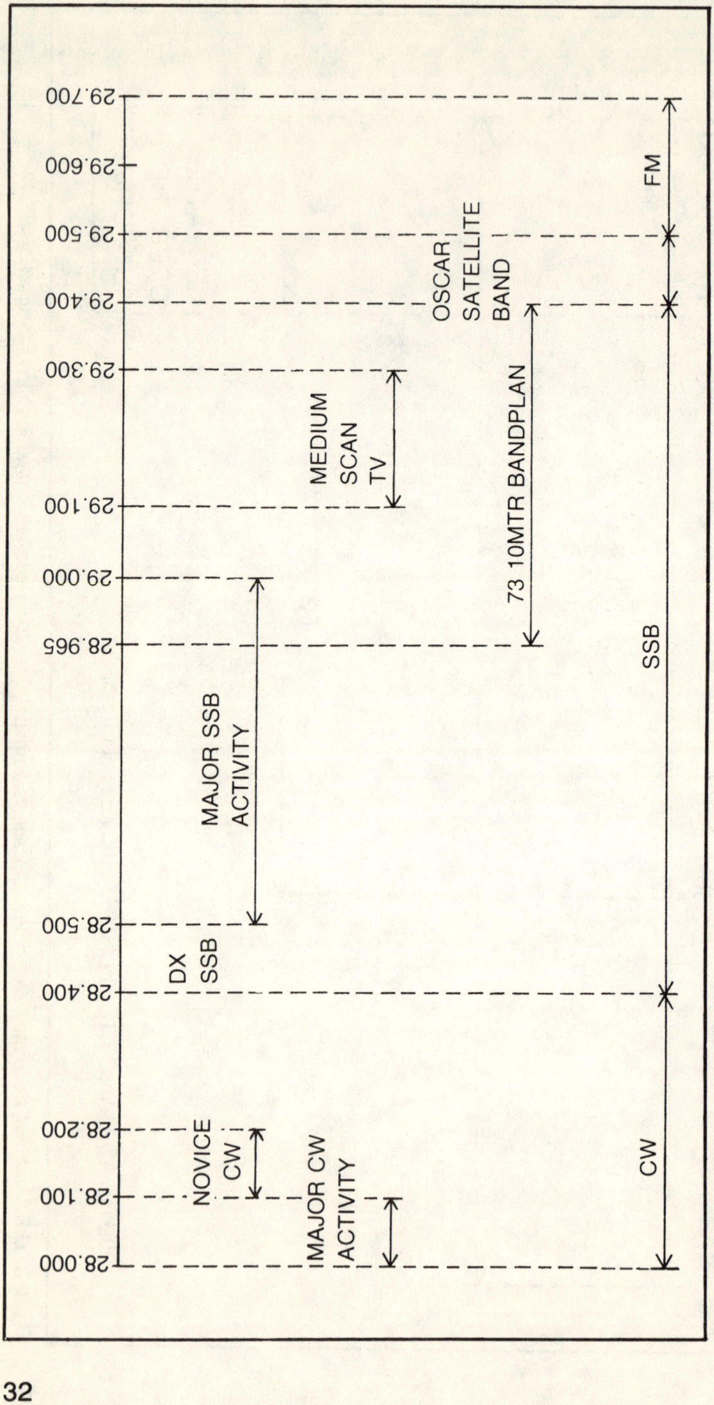

Fig. 2-9. Bandplan of 10-meter Amateur services.

Hundreds of CB antennas of every conceivable size and description are available to the 10 FM enthusiast. These arrays are easily trimmed for operation on the 29,600 -kHz band, and used antennas are quite plentiful. Additionally, may CB'ers discard antennas they have overcut when trimming them for a low SWR. Usually, such antennas can be used on 10 FM without any additional tuning!

ANALYSIS OF 10-METER SPECTRUM

Encompassing more than 1.5 MHz of RF spectrum, 10-meters is Amateur Radio's largest High-Frequency band. Each kilohertz of this outstanding band is used to its fullest, and performs a specific function. A breakdown of the Amateur services presently active on 10-meters is shown in Figs. 2-8 through 2-12. It should be emphasized that a combination of government regulations and "gentlemen's agreements" are employed to assure that all modes using 10-meters live in harmony.

CW activity on 10-meters is situated between 28,000 kHz and 28,400 kHz, with 28,000 to 28,100 kHz being the prime DXing area. The American Amateur Novice band occupies 28,100 to 28,200 kHz; novices being restricted to 250 watts input and CW operations only. Foreign Amateurs often operate AM or SSB in the 28,400 to 28,500 kHz region, however this region is not authorized for voice operation by U.S. Amateurs. AM, FM, SSB, and CW activity is authorized for the 28,500 to 29,700 kHz region, however, "gentlemen's agreements" have resulted in the following general-spectrum use.

The majority of worldwide single-sideband activity on 10 meters occupies the 28,500 to 28,800 kHz range, with AM activity

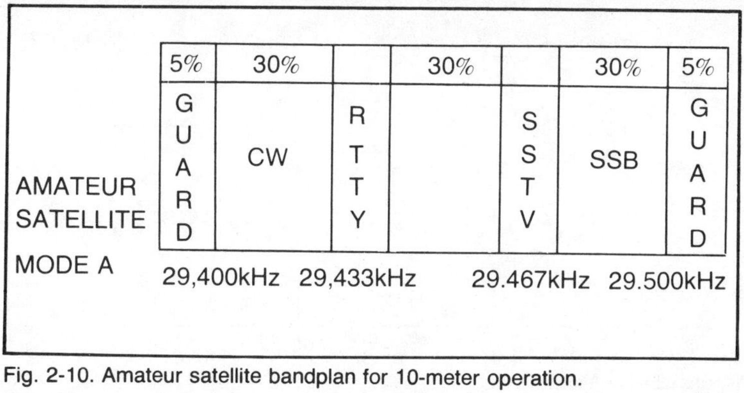

Fig. 2-10. Amateur satellite bandplan for 10-meter operation.

Channel	Freq.(MHz)	Channel	Freq.(MHz)
1	28.965...Listening & calling	21	29.215...Oscar coordination
2	28.975...Autocall monitoring	22	29.225
3	28.985...County hunting—not	23	29.255...SSTV
	rag chew	24	29.235
4	29.005...Beacon monitoring	25	29.245...Repeater
5	29.015	26	29.265...Repeater
6	29.025...Rag chewing(lowest)	27	29.275...Repeater
7	29.035	28	29.285
8	29.055	29	29.295
9	29.065	30	29.305
10	29.075	31	29.315
11	29.085	32	29.325
12	29.105	33	29.335
13	29.115	34	29.345
14	29.125	35	29.355
15	29.135	36	29.365
16	29.155	37	29.375
17	29.165	38	29.385
18	29.175	39	29.395
19	29.185...Repeater channel	40	29.405...Oscar listening
20	29.205...RTTY		

Fig. 2-11. The 73 Magazine for channelized AM operation of converted CB transceivers. This arrangement allows simple upward modification of exactly 2 MHz for operation on 10 meters.

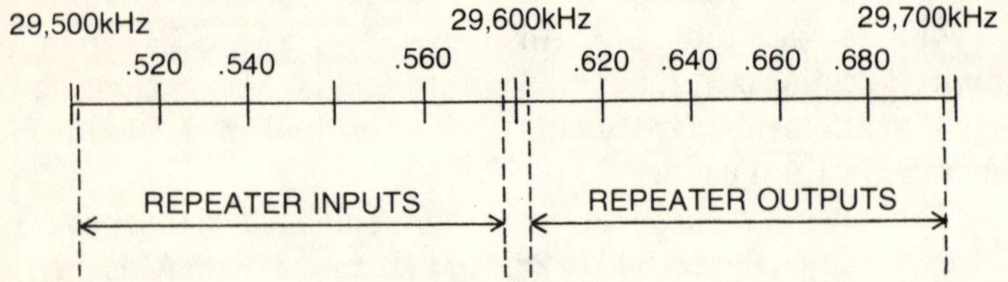

REPEATER INPUTS	INTERNATIONAL DIRECT FREQ.	REPEATER OUTPUTS
29,520 kHz	29,600 kHz	29,620 kHz
29,540 kHz	SECONDARY DIRECT FREQ'S	29,640 kHz
29,560 kHz	29,620 kHz	29,660 kHz
29,580 kHz	29,640 kHz	29,680 kHz
	29,660 kHz	

DIRECT OPERATIONS ON REPEATER OUTPUT FREQUENCIES IS ALSO PERMISSABLE.

Fig. 2-12. 10-meter FM bandplan as of mid 1979. Note 100-kHz repeater input/output spacing.

occasionally appearing near the upper range of this region. Slow-scan TV communications center around 28,680 kHz, and extend higher or lower in frequency as activity requires. Channelized AM and/or SSB activity, created primarily by the use of converted CB sets, occupy the approximate frequency range of 28,965 to 29,400 kHz (Fig. 2-11). The frequency range of approximately 29,100 to 29,300 kHz has been established as the spot for medium-scan TV and experimental slow-scan TV operation. The Amateur satellite allocations occupy approximately 29,300 to 29,500 kHz, this range being subdivided for Cw, SSB, and SSTV activity according to the specific satellite in use at a particular time. Finally, we reach the 10-meter FM area which occupies the 29,500 to 29,700 kHz range. The 10-meter FM spectrum is also subdivided into specific functions; 29,500 to 29,600 serving as repeater input frequencies while 29,600 to 29,700 serves as repeater output frequencies. Additionally, the majority of 10 FM "direct" activities centers around 29,600 kHz. Generally speaking, 10-meter FM repeater in the United States employ 100 kHz input/output spacing, thus amateurs transmitting "into" repeaters on 29,520 kHz are repeated on 29,620 kHz, etc. See Fig. 2-12.

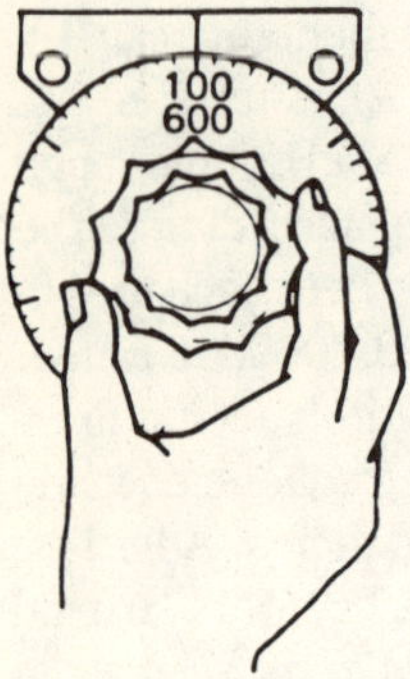

Chapter 3

Operating On 10 FM

One of the best ways to enjoy 10 FM involves letting your rig do most of the work. Since the majority of activity for specific areas usually centers on a specific frequency, a squelched transceiver can monitor that channel indefinitely. You need not be notified until activity begins to stir. While many amateurs in remote or sparcely populated areas benefit by monitoring the "international direct frequency" of 29,600 kHz, others elect to monitor output frequencies of 10-meter repeaters serving their area. Either method produces excellent results, and the ultra-enthusiast may add a simple 2, 3, or 4 channel scanner for additional flexibility (Fig. 3-1).

Mobile operations from vantage locations and mountain tops are particularly enjoyable on 10 FM. The additional dB of these occasional activities often permits low-power rigs to perform as well, or better than, base stations setups (Fig. 3-2).

Amateurs desiring to "try the water before jumping in," or operators desiring to investigate the increasing excitement on 10 FM as they read this book, can use their existing receiver or transceiver in the AM mode, and slope detect signals located in the approximate 29.6 MHz range of 10-meters. Slope detection simply means a receiver is tuned to the side rather than the center of received signals. In other words, "tune around" the FM signal until acquiring maximum reproduced intelligence. Usually, one weekend of such tuning and investigation will convince any and all

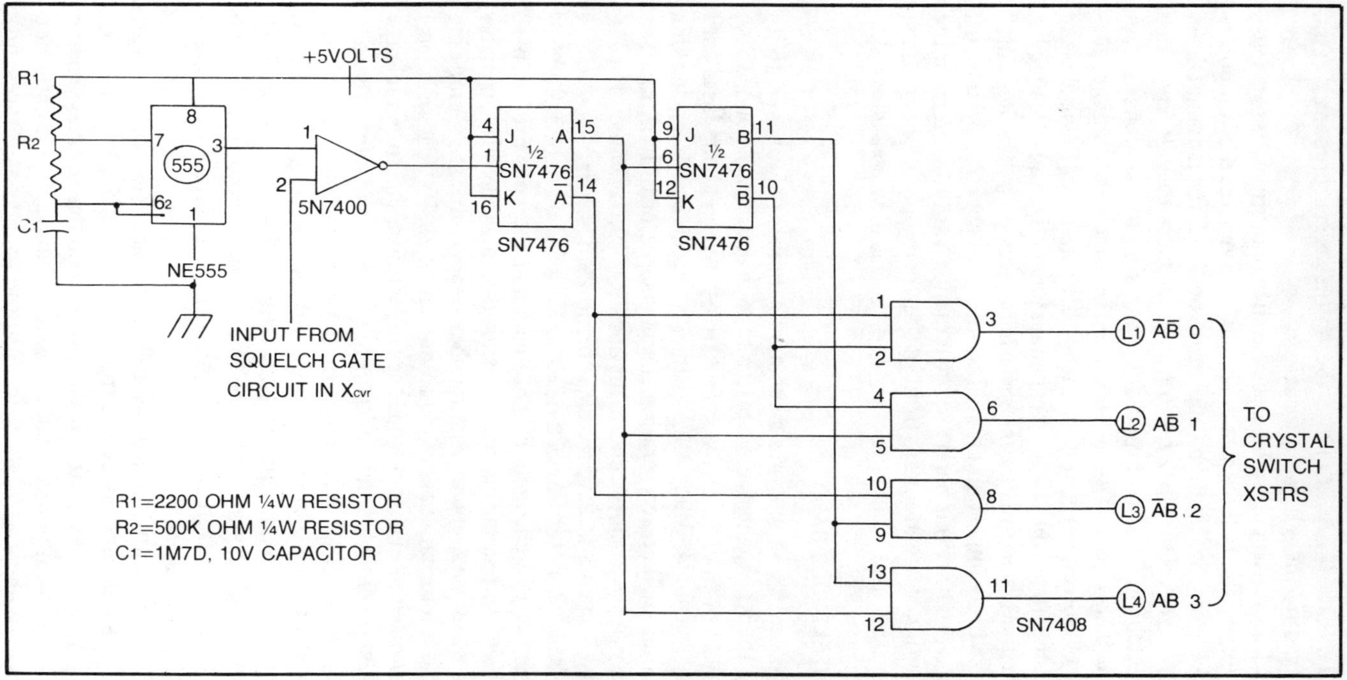

Fig. 3-1. A simple, 4-channel scanner circuit which can be added to a crystal-controlled FM transceiver. The unit sequentially turns on transistors, which then complete the FM rig's crystal-to-ground connections.

skeptics that 10 FM is one of the hottest areas presently being pioneered by today's radio amateurs.

Before delving into the joys of 10 FM operations, lets discuss some established ethics which should be considered by all potential 10 FMers. DX stations appearing on the International Direct Frequency of 29,600 kHz may or may not elect to shift frequency after establishing a work plan for contacting the stations on frequency. This decision should be based on band conditions, congeniality of on-frequency stations, available time, etc. Regardless of that decision, no single station should monopolize the frequency in use, or the DX station's time. Sharing assets on a "Golden Rule" basis is vitally important on the 10-meter FM spectrum. There's a strong possibility this band may eventually rival 160-meters for the designation of "gentlemen's band" if courteous operations continue to flourish. Such visions are, indeed, pleasant to consider.

THE WEEKENDER BAND

While 10 FM supports a substantial amount of casual activity during the week, this band's prime time is during weekends. This booming excitement begins each Friday afternoon and lasts until Sunday evenings. Many DX stations frequent 10 FM during this time, thus further accelerating the pleasures of FM-DXing. The majority of DX activity centers around 29,600 kHz, with minor DX activity occuring at 29,620, 29,640 and 29,650 kHz.

Unlike 2-meter FM activities, calling CQ on unoccupied frequencies or FM channels is quite common and very useful on 10 FM. Many times, during both weekdays and weekends, 10-meter FM is wide open, and a large number of amateurs are monitoring either the international calling frequency of 29,600 kHz or a local 10-meter repeater output frequency, awaiting action to commence. Obviously, someone must initiate this activity and alert others to band openings. 10 FM is unique in this respect, since a squelched rig doesn't produce an indication during band opening unless there are communications in progress. CQs should be kept brief on 10 FM, since other stations will either receive or not receive the calls, and to be sure that interference to possible in-progress communications are minimized.

Most of the amateurs using 10 FM operate in the 10 to 50 watt class, thus this band/mode is a true QRP haven. Unlike most QRP operations, however, 10 FM operators need not dodge weekend operations and group communications on international gathering frequencies like 29,600 kHz. Indeed, the prime desire presently

Fig. 3-2. Author Dave Ingram, K4TWJ, straps a 10-meter "Ringo" onto his car for some FM fun high above the city of Birmingham, Alabama. Note mobile 10-meter and 2-meter antennas on the car as well.

expressed by many amateurs concerns preserving the low-power communicating abilities of 10 FM.

USING REPEATERS

Operating repeaters with 10-meter input and output frequencies is quite enjoyable, provided all parties have a thorough understanding of the associated parameters and capabilities. Ground-wave coverage and short skip conditions on 10-meters provides substantially more range and a larger user audience than afforded by 2-meter FM. Consequently, 10 FM operators should be particularly conscious of their attitudes and conversation length. Simply stated, sharing of 10 FM frequencies and repeater times is decidedly more important than it is on 2-meter FM. Self-discipline and courteous operating tactics are vital. Likewise, distant stations communicating with 10 FM stations through such repeaters should be welcomed, and courteously allowed to communicate with each person as desired rather than being asked to change frequencies. It's understood that emergency preparedness and long-range communications are a prime function of 10-meter repeaters. Amateurs disapproving of this practice should restrict their operations to higher frequencies such as 2-meters and 70 cm.

Operating repeaters with 2-meter input and 10-meter output frequencies is an ultimate pleasure for FM enthusiasts. Suddenly, Star Trek visions of intercontinental communications with small, hand held "talkies" is reality for all amateurs holding General Class or higher licenses. Fantastic is a mild description of such capabilities! Obviously, the use of 2 to 10-meter repeaters requires highly considerate and disciplined tactics by all users. Likewise, strict repeater control or highly sophisticated assessing techniques should be incorporated to assure local-to-local 2-meter operations are not retransmitted on 10-meters, and to assure the repeater's 2-meter output is disabled when its 10-meter output is enabled. All 2-to-10 meter repeaters should also include a fail-safe system to prevent unauthorized or uninformed stations from accessing its 10-meter link. Since the vast majority of 2-to-10 meter repeaters have an output on the International Direct Frequency of 29,600 kHz, all users must realize the importance of frequency sharing on a world-wide basis—no exceptions. The acceptance and future of 10 FM depends on today's pioneers establishing and maintaining this concept.

The previously described parameters also hold true for remote-base setups. The primary purpose of these units is extended range, which means they usually operate on 29,600 kHz. Again, strict control and limited input range must be employed to assure only an authorized station or stations may use the machine, and that activity is restricted to specific time slots.

The possibilities of remote-base setups should not be underestimated. A rural-area amateur can interconnect his 10 FM and 2 FM transceivers by two carrier-operated relays, use 0.5 watt, and an in-shack "rubber duck" antenna on the 2-meter unit, and enjoy a 10-meter neighborhood "remote." A few metropolitan-area amateurs, experiencing criticism rather than support from local 2-meter repeater groups, can pool funds and establish their own 10-meter "dream machine." Circumstances of this nature usually dictate that the equipment be located in one of the user's homes, and requires specific time operations and continuous monitoring during on-the-air periods. Remember, the sanctity of 29,600 kHz (and all other 10-meter frequencies) must be preserved at all costs. If a system can't guarantee this, it must be removed from service. See Figs. 3-3 and 3-4.

UNDERSTANDING 10 METER PROPAGATION

Each amateur band boasts a set of conditions unique to that particular band, and 10-meters is no exception. This high-frequency band is "opened" for distant communications primarily by our sun's action on the upper ionosphere. When the ionosphere is cool (night times), 10-meter signals are lost to space. Predictable communications during these times are limited to ground wave coverage. Fortunately, however, this coverage is extremely good on 10-meters, and communication capabilities of 100 to 200 miles is

Fig. 3-3. This remote-base setup was assembled by interconnecting a two-meter FM unit and a 10-meter FM unit. Cabinet in the middle houses CORs and impedance-matching networks. Either transceiver may be removed from this setup and used separately as desired. An antenna tuner is highly desirable for tuning 28.5-MHz antennas to the 29.5-MHz part of the band.

not unusual. When the ionosphere is warm (during daytime hours), 10-meter signals are reflected. The quality of ionospheric reflection is directly influenced by the amount of ultraviolet, infrared, photon, etc. energy being received from the sun at that time. Naturally, these parameters vary hour by hour, day by day, month by month, and year by year—in 11-year sunspot cycles.

The prime times for 10-meter DX communications (assuming an established period in a specific sunspot cycle), are when the ionosphere undergoes a daily change from hot or cold at one end of the communications path while sunlight hours are experienced at the path's other end. Propagation between the U.S. east coast and Europe, for example, are quite good during approximately 1400 GMT, and 1900 GMT.

The earth's ionosphere may be hypothetically visualized as a form of variable mirror to radio signals. During night-time hours the mirror becomes a semi-clouded glass, passing signals without reflections. During early mornings, the sun-energized mirror forms with a westward progression. The leading and trailing edges of this imaginary mirror are prime long-distance aids, since an ionospheric change is taking place, and because the imaginary mirror encompasses a maximum communications range at that time. Natural and man-made phenomena may alter these conditions, however their occurence is definitely an exception rather than the rule. Comets and/or meteor showers, for example, may produce ionized trails which will reflect radio signals. During 1958, an experimental man-made ionized cloud, known as "Operation Smokepuff," was conducted in the desert southwest with positive results. Many amateurs used the cloud to communicate with other distant stations. Unless the 10-meter enthusiast has unlimited time to pursue such special-effect conditions, he will realize maximum benefits when using more conventional techniques.

The contributing factor for successful operations on 10 FM involves recognizing 10-meter opportunities and being on the right frequency when a band opening occurs. The previous discussion of 10-meter propagation will serve as a reliable guideline toward that means. Additionally, 10-meter SSB and CW activity will provide an accurate indication and direction of band opening. Whenever 10 SSB openings peak, activity on 10 FM also booms. Another reliable method of determining 10-meter SSB/CW or FM openings involves listening for specific 10-meter beacon signals. These beacons are low-power transmitters operated by energetic amateurs in various areas of the world. Their purpose is simply to

Fig. 3-4. There are a number of new all-mode, two-meter transceivers appearing in Amateur stations each day. These units may be pressed into remote-base service, along with the stations low-band transceiver, with minimum difficulty and expense. The voice-operated control relays in both units can provide instant operation while carrier-operated relays are being constructed.

alert others to 10-meter band openings. Since the number of 10-meter beacon transmitters, their exact frequencies, and power levels vary according to owner funds, operating times, etc, specific up-to-date list cannot be included in this book. Readers should consult monthly amateur magazines for the latest news of these beacons.

MEANINGFUL COMMUNICATIONS

Ten-meter FM has the capability of becoming a leading amateur band for truly purposeful amateur communications. The personalized atmosphere of FM, which has been cultivated from 2-meter activities, may serve as an illustration of how amateurs can share their lives, thoughts, and ideals with fellow amateurs in other lands.

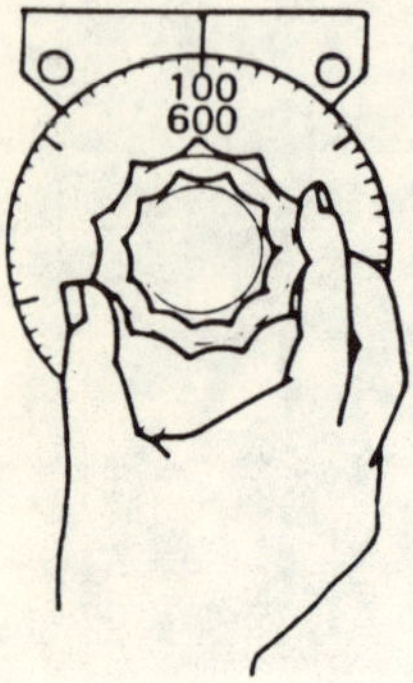

Chapter 4
10-Meter FM Equipment

During the early days of radio, amateurs necessarily constructed their own equipment from whatever items they found available and deemed feasible. Since ready-made units were relatively unknown to these communications pioneers, their implementation techniques truly established the amateur spirit which we know today. While many parameters have been altered by modern technology, today's amateur radio frontiers still reflect this Golden Age of Radio. Commercially manufactured amateur transceivers operating on 10-meter FM, for example, were relatively unknown until mid-1978. Consequently, most amateurs either home constructed such units, or modified commercial business radios for operation on 10. Occasionally, a military/commercial/amateur manufacturer (Collins Radio, for example) would include NBFM capability in a specific unit, but such options seldom achieved wide recognition. Possibly, this situation was due to propagation nulls when the gear was marketed, or, perhaps it was due to a general disinterest in massive, bulky units which didn't include such features as squelch, tone access, etc.

During the late 1970s, however, this situation began to change. Several amateur equipment manufacturers began noticing the growing number of 10 FM enthusiasts, and the tremendous low power DX capabilities of this band/mode. Today, as 10-meter FM continues to climb in popularity, amateur equipment manufacturers are recognizing the market which is evolving with 10 FM, and are creating highly outstanding units which perform magnificently on 10 FM. The similarity between this recent "initiation of 10 FM" and the late 1950s "initiation of 2 meter FM" is truly phenomenal!

As of early 1979, the majority of equipment in use on 10 FM consisted of converted low-band business radios. Such units were manufactured by Motorola, General Electric, RCA, etc. primarily for government service use in the 30 to 35 MHz range. As these two-way radios are removed from service, amateurs purchase them on a first-come, first-served surplus basis. Naturally, the number of amateurs anxious to purchase these inexpensive units substantially exceeds available supplies. Thus, "waiting lists" for interested purchasers are usually maintained by the miniscule number of amateurs able to secure these lowband FM units.

As 10-meter FM continues to grow in acceptance and popularity, we can expect to see a market which will feature a variety of outstanding equipment. If we view presently available 2-meter equipment as a guideline for this market, we can foresee 10-meter scanning transceivers with optional microprocessor remote control for programming, priority channel capabilities, front-panel controlled memories, selectable deviation control (for optional use with medium-scan TV) accessory connections for carrier-operated-relay control via a UHF link, and much more. (Fig. 4-1).

During the next few years, 10 FM's rise to fame and glory should be an outstanding phenomenon to behold. The first phase of

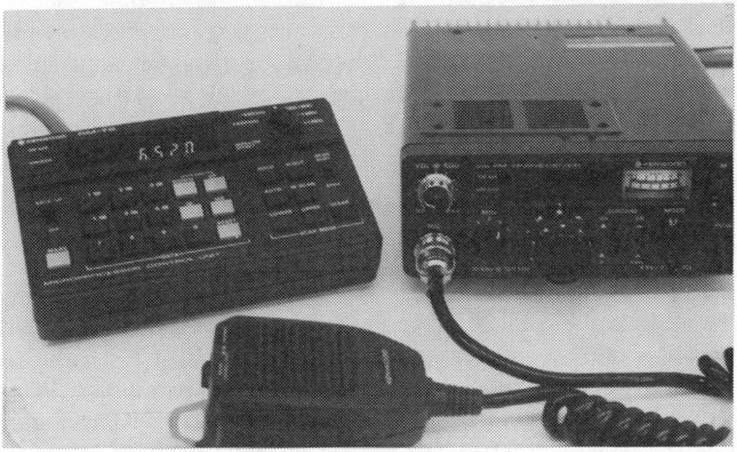

Fig. 4-1. This microprocessor-controlled 2-meter FM unit may prove to be a forerunner of 10-meter FM transceivers. The controller scans frequencies, stores them in memories, scans memories, and searches for busy or open channels.

this debut—conversion of single frequency business radios for operation on 29,600 kHz—has created excitement throughout the United States. Many amateurs, unable to secure a surplus 10 FM unit, added this capability to their existing high-frequency transceiver. Others, searching for a small 2-meter type unit, converted CB sets for use on 10 FM. Few possibilities were left uninvestigated during this enthusiastic search for equipment which was capable of 10-meter FM operation.

Phase II, the most important aspect of 10 FM's growth, was the introduction of commercially manufactured amateur equipment for the 29-MHz band. This significant breakthrough is just beginning. Companies such as Yaesu Radio (FT901DM Transceiver) and Comtronix (FM 80 Transceiver) introduced 10-meter FM equipment, and activity flourished. Practically overnight, amateurs throughout the world began appearing on the International Direct Frequency of 29,600 kHz. Activity was booming, and other channels such as 29,620, 29,640, 29,650 29,680 kHz, etc. began acquiring an increasing share of 10 FM signals. Remote-base setups, and 2-to-10 meter repeaters also began to hit the airwaves in full force, and 10 FM became a unique "weekender band" for many amateurs.

Judging by the overwhelming past record, 10-meter FM has a tremendous future in store and its opportunities should, indeed, prove endless. If you would like to share the pleasures of pioneering one of the genuine modern frontiers of amateur radio, now is the time for action. 10 FMers are experiencing the time of their lives in this area of wide open communications, gentlemanly operations, and pure excitement. The recipe for involvement is quite simple: secure a 10 FM unit, or convert an existing AM transceiver for use on FM, and join the fun. You've very little to lose and worlds of enjoyable contacts to gain!

THE YAESU FT901DM

The first major breakthrough in commercially manufactured amateur equipment for 10 FM happened when Yaesu Radio introduced their Model FT901DM unit (Fig. 4-2). This all mode, deluxe model, high-frequency transceiver includes a multitude of features which any serious radio amateur will find extremely enjoyable to use and operate. Ten-meter FM capabilities of the FT901DM are, likewise, highly outstanding. The FT901DM includes digital and analog frequency readout, with calibration accuracy within 100 Hz. Frequency memory capability permits any frequency to be stored and used for either transmit or receive functions. This feature

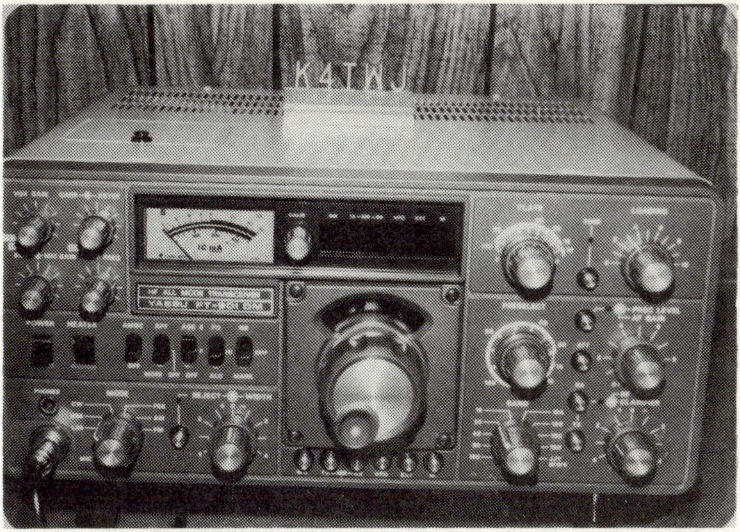

Fig. 4-2. Yaesu's FT-901DM top-of-the-line, all-mode, HF transceiver provides outstanding performance on 10-meter FM. Features include digital readout, frequency storage, AC/DC power supplies, off-air monitor, variable output-power control, and more.

allows the FT901DM to operate any 10-meter repeater "split." During simplex operations, the memory may be used to store separate frequency contacts which the operator may recall at will. A highly efficient noise blanker and squelch circuit are also provided in the FT901DM, with controls for both of these functions conveniently placed on the transceiver's front panel. During operation on 10 FM, the FT901DM's front panel meter may be switched to read relative output or final amplifier cathode current. During operation on 10 FM, the FT901DM's front panel meter may be switched to read relative output or final amplifier cathode current. A front-panel-selected 20 dB attenuator is provided to protect the receiver's "front end" from nearby high-power transmitters.

Other FT901DM features include a 10 second "tune" function which may be used independent of any front panel control, an automatic microphone-control circuit which prevents background noises from being transmitted, a self-contained 12-volt DC power supply which may be used for mobile/emergency operation, and a separate heater switch for the unit's 6146B final amplifier tubes.

The FT901DM is rated at 80 watts input on 10 FM; its typical output is 30 watts. The deviation is set at ±5 kHz. The FT901DM's outstanding receiver is rated at 0.3 microvolts for 20 dB quieting.

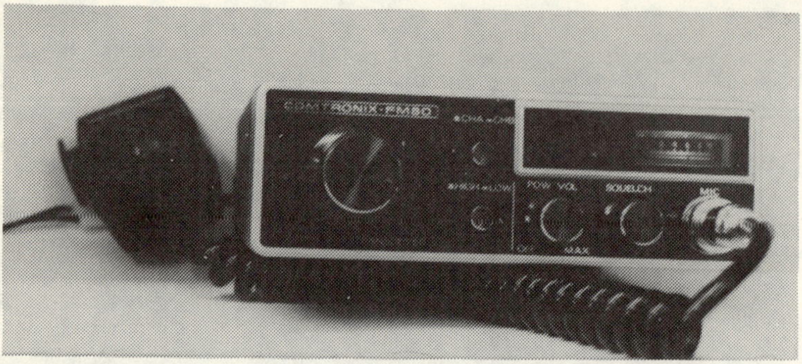

Fig. 4-3. The Comtronix FM80 10-meter FM transceiver which created widespread enthusiasm during early 1979. The compact unit operates on 80 channels, features transmit/receive indicators, separate modulation indicator, S/power meter, and repeater offset switching. It can be used fixed, mobile, or portable.

Considering the relatively high cost of the Yaesu FT901DM, it's somewhat unlikely an amateur would purchase this unit for exclusive 10 FM use. However, an amateur purchasing a new transceiver for fixed station use would surely benefit from the FT901DM's many assets—particularly its 10 FM capability. The FT901DM is distributed by *Yaesu Musen USA, Inc.*, P.O. Box 498, Paramount, California 90723.

THE COMTRONIX FM80

During the early part of 1979, Comtronix, Inc., 116 Lark Center Drive, Santa Rosa, California 95401, announced the exclusive 10 FM transceiver for amateur use. This compact and relatively inexpensive unit quickly gained widespread popularity, and 10-meter FM activity flourished (Fig. 4-3).

The Comtronix FM80 covers the frequency range of 28,910 to 29,700 kHz in 80 ten-kHz steps. This range is divided into two 40 channel sections which are selected via a front panel switch. Another front panel switch selects either high power (10 watts) or low power (1 watt) output mode. A plastic-faced front-panel escutcheon encloses the frequency/channel display, S/power meter, and two light-emitting diodes. One LED is bi-polar: green during receive and red during transmit. The other LED varies in intensity during transmissions, according to microphone-input audio. Repeater offset capability is provided by a push-pull switch mounted on the unit's squelch control. The FM80 is supplied with a matching microphone, mobile mounting bracket and fused power cord. The unit operates directly from any 12-volt DC power source capable of supplying 2.2 amperes. Transmitter bandwidth is factory set at ±5 kHz, and the FM80's receiver is rated at 0.5 microvolt for 20 dB quieting.

The extreme flexibility of the Comtronix unit makes it a true pleasure to own and operate. The 40-position frequency/channel switch permits instantaneous and accurate frequency changes or band monitoring, while the combination of an RF-power meter and modulation-indicating LED assures full monitoring of the transmitted signal. The unit is equally suited for mobile, portable, or fixed-station use, and the author's unit has extensively proven its ability during 10 FM DX activities. The fully solid-state design, coupled with broadband circuits allows "instant-on" and carefree operations as desired.

Mobile operations on the high-frequency amateur bands have undergone many changes during recent years. Today's crowded highways and byways, coupled with hectic pace of living, have almost eliminated single sideband mobile activities. Tuning such rigs while an auto is in motion is simply too dangerous to be enjoyable. Channelized operations with squelch-muted rigs are a logical and safe solution to this dilemma. The Comtronix unit provides this ability quite efficiently, and it can be enjoyed during any available opportunity.

CONVERTING THE MOCOM 10

The Mocom 10 is a compact, 2-way business radio manufactured by Motorola for use in the approximate frequency range of 30

Fig. 4-4. This Motorola Mocom 10 has been converted to a 10-meter FM transceiver for Amateur use on 29,600 kHz. The original faceplate was reversed, painted flat black, and marked with decals. Knobs are miniature types from Drake, and the microphone is a Telex "Double Leader."

Fig. 4-5. Top view of the Mocom reveals a compact unit which is easily serviced. The 3 screws holding the speaker can be removed, allowing the speaker to be folded back out of the way for access to the unit. The PC board adjacent to the speaker is a tone-encoder unit.

to 35 MHz. This fully solid-state unit delivers approximately 10 watts output on a single frequency selected by its user. The Mocom 10 is approximately the size of any popular 2-meter FM mobile unit, operates directly from a 12-volt DC source, and its 2-piece case is heavy duty plastic. A relatively large number of these units have been removed from service during recent years, and they are now being made available for amateur use. Many large surplus outlets handle Mocom 10s on an "as available" basis, or the energetic amateur may spot a local company planning to remove their Mocoms in the near future. There are several reasons why the Mocom has gained widespread acceptance and use on 10 FM. The units are small, state of the art, and they are usually inexpensive. Their design is basically simple and straightforward. Finally, the conversion on a Mocom to 10 FM merely consists of replacing transmit/receive crystals, and retuning the affected stages (Fig. 4-4).

Assuming the 10 FM enthusiast acquires a Mocom 10, the first step should involve checking the unit to assure everything is intact and operating properly. Loosen the two rear screws and slide the unit from its case. Check that there are no burns or obvious arcs on the P C board and/or parts (see Fig. 4-6). Also check to ascertain both transmit and receive crystals are firmly seated in their sockets. With the unit placed speaker-up and front facing you, the transmit crystal will be near the left rear (Fig. 4-5). Next, apply power to the Mocom and check its operation. There are four wires connected to this unit's power plug: three positives and a ground. One positive is for the unit, one is for ignition switch/relay control, and one is for Private Line® tone. Connect all three of these together for one positive lead. Apply 12 volts between the positive and negative leads, and switch on the unit. (Watch for possible smoke, and ascertain the unit isn't hung up in the transmit mode!) Receiver operation, including squelch action, can be checked with the use of a signal generator tuned to the Mocom's receive frequency. An obvious squelch-break and quieting action will indicate proper operation. The transmitter can be checked by using a dummy load/wattmeter and a general coverage receiver. This is also a good time to note output power and frequency of operation.

Fig. 4-6. PC board view of Motorola's Mocom, showing various holes for alignment. Smaller holes accept prods from test meter.

Following a successful checkout of the Mocom in its original form, its crystal frequency calculations can be reviewed and new crystals ordered. Using a supplied Mocom manual, calculation *for the appropriate frequency range* are compared with the unit's installed crystals. Assuming, for example, a unit transmits on 30.999 MHz, the transmit crystal will be 10.333 mHz (one-third the output frequency). The receive crystal's frequency is determined by the formula: receive frequency—455 kHz/8. Thus 30.999—.455 = 30.544/8, or 3.818 MHz. This formula must be altered for use below 30 MHz, however, and the new formula is: receive frequency—455 kHz/7. In order to receive 29.600 MHz, for example, the calculation would be 29.600—.455 = 29.145/8, or 4.163 MHz. As previously mentioned, the Mocom's instruction manual should be understood thoroughly while making these calculations. Then, a crystal order can be placed with a reputable manufacturer. Any attempt to cut costs by purchasing off-brand crystals is definitely not advisable. Don't skimp on crystals!

The Mocom can be given a liberal refurbishing during the anxious wait for crystal arrival. A coat of paint to match its proposed surroundings (auto dash, etc.), and modern knobs can produce a refreshing facelift for the small unit. Additionally, the Mocom's front panel can be removed, reversed and painted to match the "new rig".

After the new crystals are acquired and installed, the unit should be retuned for amateur-band operation. Explicit alignment instructions are included in the Mocom manual, however, a brief summary will now be presented for amateurs unable to secure or understand that information. Fortunately, the Mocom's printed circuit board is fully marked for all adjustment and metering points. This consideration greatly simplifies tuning procedures. Separate circuits are used for the transmitter and receiver sections, further assisting the amateur converting this unit.

The receiver should be aligned first, since it may then be used to monitor on-frequency activity during the final stages of transmitter alignment. A suitable signal source may be a signal generator tuned to the Mocom's 10 FM frequency, and connected to the unit's antenna jack. Always use a minimum acceptable generator output level. Alternatively, the station's 10-meter CW/SSB transceiver may be used as a signal source. If this concept is employed, the CW/SSB transceiver should be connected to a suitable dummy load and an extremely low RF output level should be maintained. An antenna should not be connected to the Mocom, and overloading

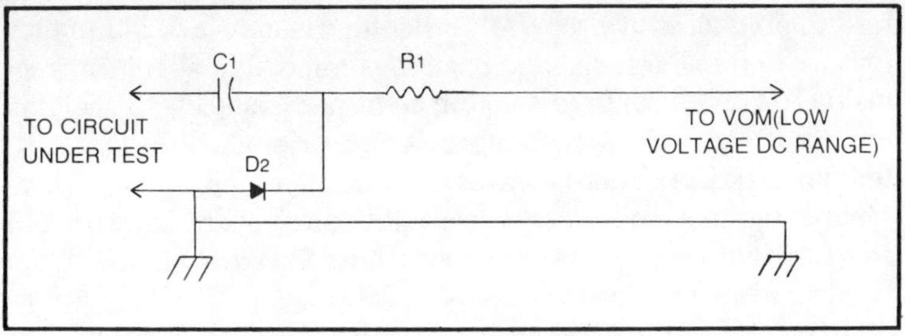

Fig. 4-7. An RF probe. Keep the probe input leads short. C1 is 0.001 uF disk capacitor, 600 V; R1 is 2.2 megohm, ½-watt, 10%; D1 is a 1N914 diode, or equivalent.

possibilities must be avoided. An RF probe and volt-ohm meter are also required for receiver alignment. Practically any 20,000-ohm per volt or higher meter, set on its 1.5 or 2.5 DC volt range, may be used, and a simple home-constructed RF probe, similar to that shown in Fig. 4-7, will perform successfully. Vary the signal source over the desired frequency range and try to get a reaction from the Mocom's receiver. Since the crystal's operating frequency hasn't been adjusted at this time, some "signal searching" may be necessary. After the signal source has been roughly located, proceed as follows: Place the RF probe's input lead in jack Ml, and its ground connection to the chassis. Tune T1, L20, and T2 for maximum indication on the VOM. This indication will be approximately 0.5 volt. Next, tune L9, L8, and L7 for maximum indication on the VOM (probe still in Ml). Then move the RF probe's input lead to M4 and tune *in this order*, L3, L4, L3, L2, L1, L6 and L5, for maximum indication on the VOM. This indication will be approximately 1.25 volt. Finally, tune the receiver on frequency, using the crystal-warping capacitor (C18) and repeat the previous adjustments to ensure maximum receiver sensitivity. Assuming the 10 FM band is open, you should now receive amateur signals when an antenna is connected to the Mocom.

Alignment of the Mocom's transmitter section is relatively simple and straightforward, provided a logical sequence like that applied to the receiver section is followed. A dummy load and in-line watt-meter should be connected to the unit's RF output, and transmitter-on time should be held to a minimum until resonance is achieved. The station's 10-meter CW/SSB transceiver may be used to monitor this tune-up progress, however its antenna should be disconnected to avoid overload problems. Remove the VOM's RF probe (used during receiver alignment) and connect the *positive* lead to ground. Move the VOM's *negative* lead to jack M3 (meter still on low DC-volt range). Place a non-metallic alignment tool inside L14 and assure the slug can be turned easily. While holding the alignment tool and watching the VOM, briefly key the transmitter and peak L14. Then release the microphone's push-to-talk and remove the tool from L14. Perform this same operation with L15 and L16, allowing the transmitter a brief "recovery period" between carrier-on times.

Next, move the VOM's negative lead to chassis ground and place the positive lead in jack M5. Using an efficient procedure similar to the previously described technique, tune T4 for maximum indication. The RF wattmeter should begin to show output at this time. Now move the VOM's positive lead to M7 and adjust L17 for maximum indication. Finally, tune C100 and C101 for an RF output of 7 or 8 watts. The transmitter may now be placed exactly on-frequency with the warping coil, L13. Be sure the speaker is in place while making this adjustment. Repeat the above adjustments briefly to insure detuning didn't take place. Remember: be ready to watch the meter and tune the transmitter before keying the mike. Don't leave the off-resonance unit keyed while hunting slugs, etc. The transmitter's mike-gain/deviation-level may be adjusted with R78, if necessary. The station's CW/SSB transceiver may be used to check the final results and audio quality of the Mocom transmitter, if desired. If the unit is low on audio output, a hand-held "power mike" will prove quite beneficial. High-quality units, such as the Telex microphones, have been proven outstanding performers on FM.

CONVERTING CB UNITS TO 10 FM

An unlimited supply of potential 10-meter FM transceivers exist in the used CB market. These compact units are quite efficient, and their cost is usually very reasonable. The 3 or 4 watt output of a converted CB set provides ample power for local or DX communication, and the 10-kHz channel spacing is ideally suited for 10 FM operations.

Frequency conversion of older tube-type CB sets using plug-in crystals is one of the quickest and easiest means of getting started on 10 FM. E. F. Johnson Company was a popular manufacturer of such units. Frequency conversion of 23 channel "synthesized" CB sets requires slightly more consideration and planning. However, the conversion cost is approximately the same as that of tube-type units. Conversion of CB sets utilizing phase-locked-loop circuits for frequency determination are more difficult to understand and convert to 10 FM. This project should not be undertaken by the inexperienced amateur. The 10 FM newcomer can enjoy a lot of operating while being restricted to between four and six channels of operation, thus a minimum number of crystals need to be exchanged in the CB unit. An inexpensive CB, modified for 10FM is shown in Fig. 4-8.

Modification of a CB set for operation on 10 FM consists of three basic steps:

- Convert the selected unit to operate approximately 2.615 MHz higher in frequency. This will permit 29,580 kHz through 29,580 kHz through 29,620 kHz operation on channel 1 through 4 of the CB set. A basic transmitter and receiver alignment will be necessary after the crystal or crystals for this range are installed.
- Replace the AM modulation system with a simple varactor FM modulator. This step consists of disabling the transmitter's speech amplifier stages, bypassing the modulation transformer and applying audio variations to a varicap circuit connected to the transmitter's oscillator crystal.
- Replace the CB unit's AM detector with an FM detector. The three previously described steps are a general out-

Fig. 4-8. Bearing a close resemblance to its original 11-meter look, this CB unit has been modified for 10 FM. Its grubby appearance has thus far discouraged theft.

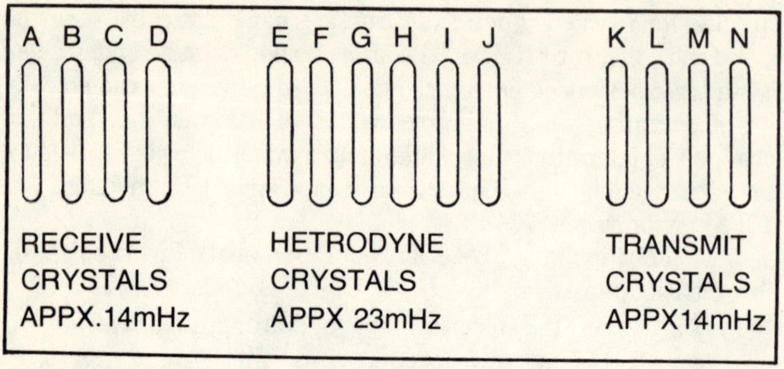

Fig. 4-9. A common 4-6-4 crystal arrangement used in many popular CB transceivers.

line for modifying CB units for operation on 10 FM. A more detailed guideline will now be presented. Since a wide variety of frequency determining techniques are used in CB sets, however, the reader may be required to handle his particular unit in a specialized manner.

As previously mentioned, early CB sets which used crystal controlled receivers and transmitters are easily converted to the 29,500 kHz range. The new crystals are merely ordered from a reputable manufacturer, and installed in their desired channel slot. The transmitter and receiver are then realigned for optimum performance at the new frequency range. When ordering crystals, full information concerning the particular CB set and its crystals should be included for the crystal manufacturer's reference.

Frequency conversion of a synthesized CB set should begin with a Sam's Photofact and Trouble-shooting guide for that particular unit. This information should be accurately analyzed until a thorough understanding of its frequency-synthesis method is realized. One of the most popular frequency synthesizing schemes employs banks of 4, 6, and 4 crystals, as shown in Fig. 4-9. Each hetrodyne crystal is beat with a receive crystal to produce 23 channel capability. As an example, crystal E will beat with crystal A through D for coverage of Channel 1 through Channel 4. Crystal F will beat with crystal A through D for coverage of Channel 5 through Channel 8, etc. During transmit, crystals E through J beat with crystals K through N to produce output on the proper channel frequency. Changing crystal E in this particular case alters both transmit and receive frequencies.

A further expansion of a hypothetical frequency synthesizer illustrating crystal placement and FM modifications is shown in

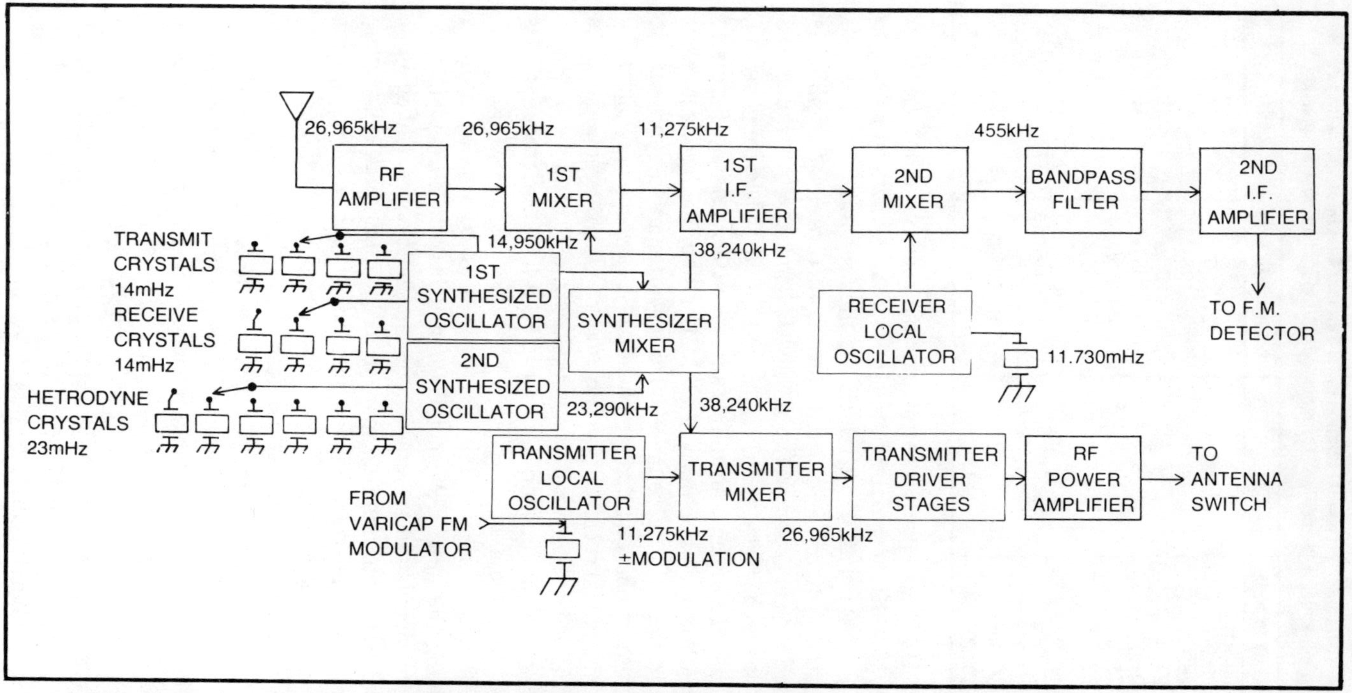

Fig. 4-10. A block diagram of a typical 23-channel CB set with frequency synthesizer as described in text. Frequencies shown are for operation on CB channel 1. Circuit example is similar to the Sharp CB800.

Table 4-1. Frequency/channel relation of CB Service. This information may be used to construct a "bandplan" for modified CB units used on 10-meter FM.

Frequency In MHz	Channel	Frequency In MHz	Channel
26.965	1	27.115	13
26.975	2	27.125	14
26.985	3	27.135	15
27.005	4	27.155	16
27.015	5	27.165	17
27.025	6	27.175	18
27.035	7	27.185	19
27.055	8	27.205	20
27.065	9	27.215	21
27.075	10	27.225	22
27.085	11	27.255	23
27.105	12		

Fig. 4-10. Frequencies illustrated in this example afford operation on 23,290-kHz hetrodyne oscillator crystal is changed to 25,905 kHz (23,290 plus 2,615 kHz), the synthesized mixer's output will be shifted to 40,855 kHz. Assuming the receiver's RF amplifier is retuned for operation on the 29.6 MHz band, the first mixer will then down-convert 10 FM signals to the first I-F frequency of 11.275 MHz. If the transmitter's mixer & driver stages are also retuned for operation in the 29.6-MHz range, the frequency shift modification will be complete. Do not attempt to short-cut the realignment phase of this modification. A *complete* realignment of transmitter and receiver is vitally important to overall results. As an additional assistance for amateurs desiring to establish a custom bandplan for CB set operation on 10 FM. A list of CB frequencies is shown in Table 4-1.

The next step in this CB to 10 FM conversion involves disabling the unit's AM modulator and installing a simple Frequency Modulation system. The unit's Photofact should again be consulted to determine the appropriate location which will prevent signal variations from affecting the modulation transformer *during the transmit mode*. Usually, this is accomplished by disconnecting one wire on the modulation transformer's primary. A basic Varicap-modulator circuit, which is directly coupled to one stage of audio amplification, is then connected to the transmitter's local-oscillator crystal. A suitable modulator circuit is shown in Fig. 4-11. The parts consisting of R5, R6, C3, C4, C5, and the Varicap should be securely mounted on a small, metal-enclosed vector board and placed near the transmitter's local-oscillator crystal. The other parts, which comprise the audio-amplification stage, may be mounted remote to this location.

Basically, the FM modulator circuit of Fig. 4-11 causes an audio signal of approximately 1 volt to be superimposed on a DC level of approximately 5 volts. This varying level changes the Varicap's bias and internal depletion-area capacitance, which in turn changes crystal-circuit capacitance and instantaneous carrier frequency at an audio rate. Variable resistor R6 serves the double purpose of isolating the audio amplifier stage from the Varicap and crystal and acting as a deviation control. Referring again to Fig. 4-11, adjust C4 to return the crystal to its proper frequency after being loaded by the Varicap circuit. R6, in conjunction with R3, is carefully adjusted to provide the desired amount of frequency deviation. This adjustment must be monitored on an auxillary receiver and/or deviation meter, and final results should be double-checked after the modified CB is reinstalled in its case. As a "quick and easy" alternative FM system, a high quality CB power mike may be substituted for the single audio stage in Fig. 4-11. If this technique is employed, the power mike's output capacitor, which blocks DC from being applied to a unit's input, should be removed. This capacitor is usually located on the mike's output terminal, near the volume control.

Several different varicaps plus some general purpose germanium diodes have been used in this circuit, and acceptable results were consistently realized.

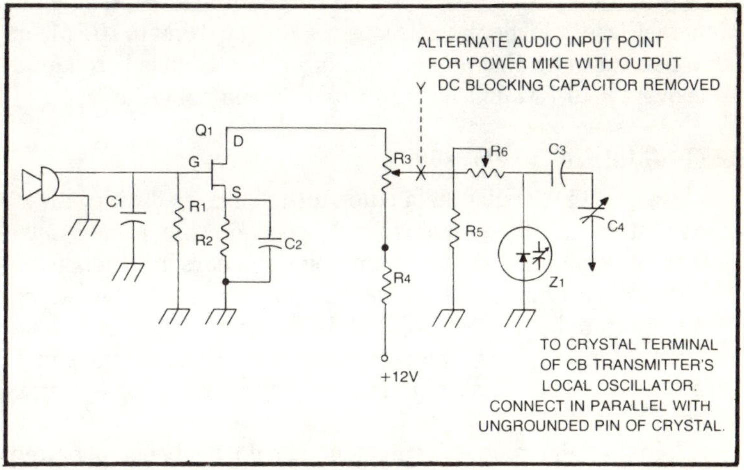

Fig. 4-11. FM modulator which may be added to a CB set: C1—0.001 uF disk ceramic; C2—5 uF, 15 V electrolytic; C3—20 pF disk ceramic; C4—50 pF disk ceramic; Q1—2N5484 FET; Z1—MV1632 Varicap diode; R1—1.5 megohm; R2—3300 ohm; R3—5,000 control; R4—1000 ohm; R5—33 k-ohm; R6—100 k-ohm control; All resistors are ¼ watt, 10% tolerance.

Finally, the CB unit's AM detector is removed, and an FM detector is directly substituted for that circuit. A highly efficient FM detector which may be used for this purpose is the ratio detector shown in Fig. 4-12. The I-F transformer should be a miniature FM unit, tuned to the CB's 2nd I-F frequency (usually 455 kHz), and including a discriminator winding on its tapped secondary. Diodes 1 and 2 should be high quality germanium types, such as 1N60, 1N270, etc. All parts for the ratio detector should be securely mounted on a small vector board and placed in its appropriate location inside the CB set. Usually, the CB unit's volume control may be used for R4. However, it should be bypassed with capacitor C7 if this option is used. While a Foster-Seeley discriminator could be used in lieu of the ratio detector, it would require an additional I-F stage which also functions as a limiter. The ratio detector's RC circuit consisting of R1, R2, and C6 has a time constant of approximately 1/20 second. Noise pulses appearing across this circuit combination do not affect the *average* values (audio) across C6, thus an additional limiter stage is not required.

The previous information is presented as a general outline for converting CB sets to 10 FM. While numerous design variations between different manufactured units will be realized, this basic information should provide a positive guideline which will accomplish the desired final results.

While any technically competent radio amateur holding a General Class or higher license may convert a CB unit to 10 FM, an FCC Second Class Radiotelephone, or higher, license is required to convert a 10 FM unit back to Citizen Band operation.

SETTING UP THE 10 FM MOBILE

Many of the popular installation techniques used with Citizen Band radios has created misconceptions of true professional mobile setups. Consequently, numerous amateurs have adopted a "cigarette lighter plug and magnetic mount antenna" habit, which leaves much to be desired. While this concept is quite acceptable for rental-car activities, the serious mobile operator can realize many additional benefits from an thoroughly planned mobile installation.

Initially, the mobile equipment should be bench-operated from the amateur's fixed location, and its on-the-air operation confirmed. This is an opportune time to fully check all equipment parameters, plus set RF frequencies and carrier deviation. This procedure will reduce the under-dash fumbles, which somehow happen practically every time an amateur begins installing mobile gear.

Realizing that permanently installed mobile equipment will be subjected to temperature extremes during Summer and Winter months, the amateur should strive for quality crystals, microphones, and relays to avoid later pitfalls. Likewise, a mobile mount which would allow the equipment to be removed at will should be considered.

High power units and surplus equipment requiring over 4 amperes of current should receive their power directly from the auto's battery rather than an under-dash connection. This battery cable should be number 12 or larger wire to avoid voltage drop or excessive heat buildup. If a wire is connected directly to the auto battery, an additional under-hood relay is suggested for on/off switching. This relay should receive power from ignition-switched voltage to prevent equipment operation when the auto's motor is off. Don't forget the essential fuse, mounted close to the battery, and wired ahead of all other circuit elements. Use a heavy-duty fuse block made for mobile, high-current service. In-line, pig-tail types are marginal for high-amperage use.

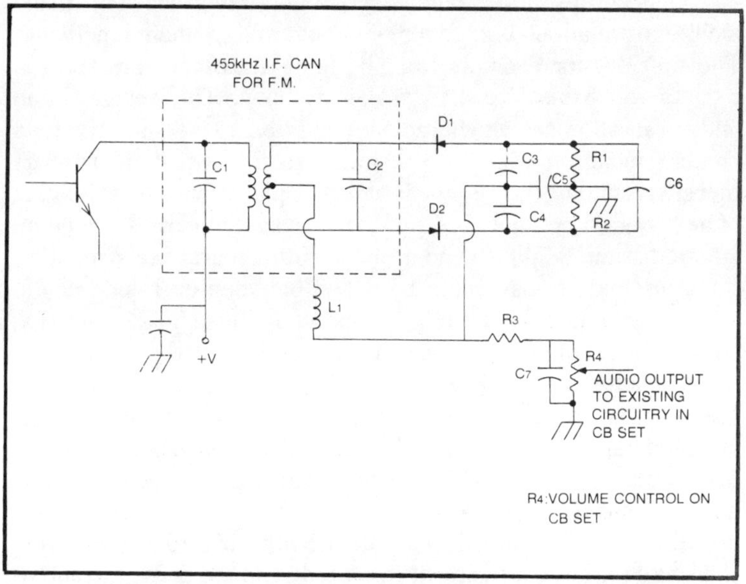

Fig. 4-12. Ratio detector which may be added to a CB unit for 10 FM: D1, D2—1N914 or equiv.; C3, C4—330 pF disk ceramic; C6—10 uF, 25 V electrolytic; C7—330 pF disk ceramic; R1, R2, R3—5600 ohms, ¼ watt, 10%; L1 is part of a 455-kHz I-F transformer.

Low power units requiring less than 4 amperes of current can be connected to an usused "accessory," terminal of the auto's under-dash fuse block. The selected terminal should be ignition-key switched. This will serve two distinct purposes: The auto's key must be used to allow operation of the equipment, and the unit will be disconnected from its power source while the car is being started. An inductive kick of very high voltages is produced when an auto's starter is disengaged after each use. This voltage spike can severely damage sensitive transistor circuits, thus ignition-switch protection is an important consideration. Automobile manufacturers overcome this problem by including an "accessory disconnect" in auto ignition switches. This feature is easily demonstrated by noting how an auto radio is de-energized when the car's ignition switch is in "start" position. Mobile amateur equipment can also take advantage of this feature by merely using ignition-switched voltage.

When routing coaxial cable from the auto's dashboard to the trunk area, try using the rider's side (right side) for convenience. Surprisingly, many installers route cable on the driver's side, which is usually blocked due to the steering wheel and floor pedals.

One of the more common problems experienced in many mobile installations is the ineffective overall grounding methods. The tailpipe, for example, may be hung via rubber or asbestos mounts which insulate it from the car body. This results in an under-car antenna which radiates ignition noise directly to a bumper mounted whip. The solution to this situation involves merely grounding the tailpipe in at least 3 places with short lengths of heavy braid or shield. One of these grounding straps should be situated at the tailpipe's end near the auto's rear. Likewise, additional ground straps should be added between each side of the auto's engine and the car body. Continuing these noise-reduction steps further, the amateur may use a transistor AM radio with its internal antenna to seek noise-radiating fenders, hood, trunk, etc. Many times, manufacturers paint these items before installation, and then bolt them into place after the paint dries. Thus, a hood or fender may be insulated from the actual body, and radiate such noise. While 10 FM is theoretically immune to noise pickup, such noises can substantially desense the I-F strip of a 10 FM receiver. If an auto's trunk lid's is paint-insulated from the body, the antenna's ground system is severely affected. One simple cure for this situation is securing metal strap or braid to the trunk lid's inside with sheet metal screws, and securely connecting this braid to the auto's body. Similar techniques may also be applied to an auto's hood, if desired.

If a trunk-lid mounted antenna is used, care must be taken to ensure that a ground connection is secured to the trunk lid and the car body. After completing antenna installation, all areas in question may be checked with a simple continuity tester or Volt-Ohm Meter. Finally, tune the antenna for a 1 to 1 SWR, and enjoy your mobile setup.

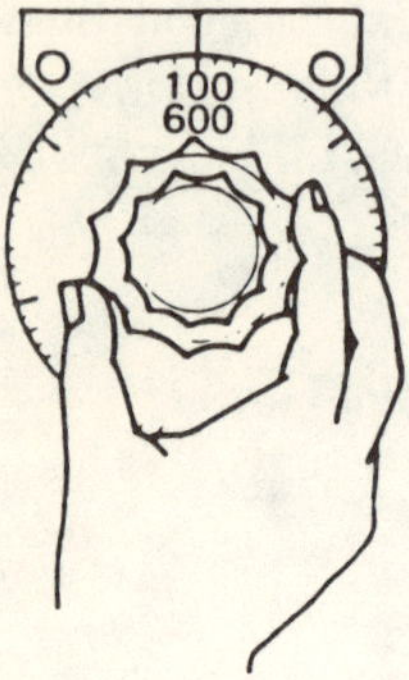

Chapter 5 Remote Bases And Repeaters On 10 FM

The full impact of 10-meter FM is best realized when the unlimited capabilities of 10-meter remote bases and repeaters are considered from a positive operational standpoint. The DX capabilities of these super-machines permits nationwide or worldwide communications on a predictable and reliable basis. The user or users, however, merely operates a compact transceiver rather than using a room full of equipment. Three or four amateurs living in rural areas may justifiably consider a 2-meter repeater unwarranted, yet this same group can have the time of their life operating a 2-to-10 meter, or 10-to-10 meter, repeater. If each member of this hypothetical group owned a 2-meter rig, a single, group-financed 10 FM unit could be linked with an extra 2-meter transceiver, and the resultant repeater/remote base placed at one member's home for all to use on a shared basis. Our small, hypothetical group would thus find their "neighborhood repeater" welcoming more guests, more DX and creating more excitement than any 2-meter repeater in the vicinity possibly could produce. If each member of this same group didn't own a 2 meter transceiver, they could convert several CB sets or business radios to 10 FM, set up a 10-meter input/10-meter output repeater, and enjoy comparable results at minimum expense.

If an individual amateur in a large metropolitan area decided to expand his horizons to include a 10 FM remote/repeater, and he couldn't kindle local interest or support, he could pursue the route alone at minimum cost. Once his simple system (which hypothetically operated during special times, and had its input on a discrete frequency known only to that operator) gained acceptance, others could join and provide financial aid for future expansions.

All that glitters cannot be defined as gold, thus some very important considerations affecting all 10-meter relay systems will now be discussed. The importance of these considerations cannot be over emphasized. Anyone planning to operate any form of 10 FM relay station should carefully study this information.

CONTROL

Strict control over 10 FM relayed signals should be maintained at *all cost*. A *fail-safe system must be employed to ensure only authorized stations can access the 10 meter link*. There should be no possible way for the 10-meter transmitter to be activated when the control operator isn't available to oversee on-the-air activities. The reasoning behind this consideration is three-fold. If the 10 meter link is left activated for extended periods, serious interference problems will be created to other 10-meter FMers. Additionally, the continuous operations of 10 FM can over-tax relay-system transmitters and ruin final-amplifier stages. Finally, the possibility of 2 meter Technician-Class amateurs accidentally gaining 10 meter access must be avoided.

LEGAL CONSIDERATIONS

Numerous decrees, regulations, and amendments affecting amateur radio are handed down from the Federal Communications Commission each month and each year. Staying abreast of these legalities requires a reasonable amount of time, yet it must be done through some conventional means. Many repeater groups elect one specific person to oversee this duty. Others go by word-of-mouth, which may or may not prove accurate and worthwhile. One logical solution, which may prove a drastic overkill when considered by pliable amateurs, is simply "when in doubt, don't." Basically, however, current FCC regulations (facts), and local ideas (opinions) should be considered before placing a 10 FM remote/repeater in operation.

As of early 1979, crossband operation of repeaters and remote-base units is legal and quite popular on 10 FM. The ratio of 10 meter input/output repeaters to 2-meter or 70-cm input/10-meter output repeaters is approximately 25/75 on the U.S. West coast. This ratio is somewhat lower on the East coast. Remote-

Fig. 5-1. A simple, easy to assemble remote-base setup using 10- and 2-meter FM transceivers, interconnected via rear-mounted cables. The cables can quickly be removed for independent use of either unit. The unit below the Comtronix 10-meter rig is a control box containing CORs, decoders, and impedance matching networks.

base setups are commonplace throughout the United States, with new units appearing or disappearing on a daily basis. Considering the vast flexibilities of remote-base setups, we realize that an accurate account of these systems is almost impossible.

QUICK AND EASY 10 FM REMOTE BASE

One of the most exciting and versatile systems which may be used on 10-meter FM is the remote-base setup. An undetermined number of these systems are presently operational on 10 FM, and their owners enjoy fantastic results. The 10 FM remote base is usually considered as a limited-access system which is used and maintained by one, or possibly two, amateurs. Two-meter-to-10-meter repeaters may embrace similar functions, however these systems are usually used and maintained by a relatively large group of amateurs. A simple, yet effective, remote-base setup which may be achieved with minimal effort will now be described. Amateurs interested in trying remote base systems should find these ideas beneficial in attempting their "first big venture" in 10 FM. See Fig. 5-1.

The concept of this setup involves interconnecting one's existing equipment to form a remote base of limited range and access. Since many amateurs possess a 2-meter FM transceiver plus a 2-meter FM hand-held unit, these items may be used with the station's 10 FM transceiver to accomplish remote-base capabilities. The 10 meter and 2 meter FM transceivers are interconnected as illustrated in Fig. 5-2, and a portable 2-meter unit is used to access this setup. Again we emphasize that strict control of the 10-meter link and constant monitoring of the particular 2 meter frequency utilized in this setup must be maintained if this very basic remote base doesn't include timers, identifiers, and remote shutdown features. The remote-base owner and operator must also stay within a couple of minutes distance of the setup, and assure his signal alone can be received by the 2-meter transceiver. Provided one doesn't live in a highly populated apartment complex, reduced-size, indoor 2-meter antennas and miniscule 2-meter power usually assure this provision.

While it may seem that a remote base of this nature could be more restrictive than enjoyable, such is not the case. This setup,

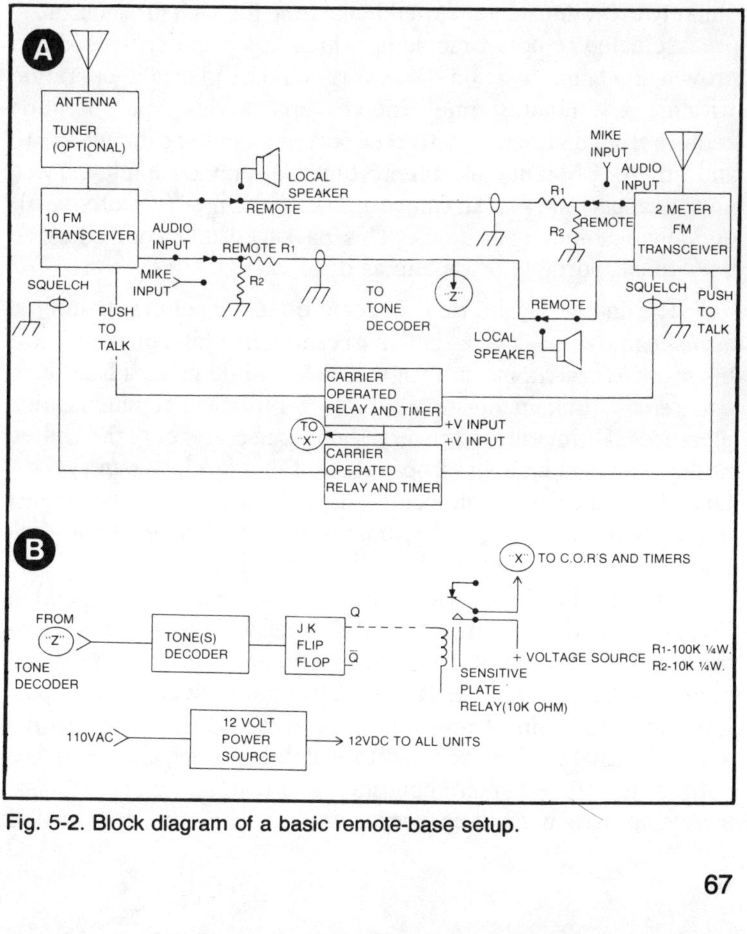

Fig. 5-2. Block diagram of a basic remote-base setup.

for example, allows an amateur to relax by his patio or pool while using his hand-held unit to contact distant amateurs. Rather than moving 10-meter FM equipment and its paraphernalia of power supplies, connecting lines, and antenna cable into the living area or den for a weekend pleasure break, an amateur merely keeps a small-hand held unit by his favorite chair as desired. Likewise, amateurs working about their house can find the occasional beckoning calls from a remote-linked 2-meter portable unit quite enjoyable to monitor.

A limited-access remote-base setup is not confined to the previously described operations. It may also be moved and used for portable or mobile activities, if desired. An amateur enjoying a day on the beach or fishing from a river bank, for example, can use his portable 2-meter gear for remote operation of the nearby remote-base setup. While 10-meter FM equipment could possibly be set up for conventional operations at a river bank or beach site, it's highly unlikely the radio amateur would take time for such procedures. A preassembled remote base setup which was previously used and proven at a home location, however, could be placed in operation within a few minutes time. The concept for this type operation centers around a home constructed portable cabinet to house the 10 and 2 meter FM units, an antenna tuner, a common ample-service power supply, and a small control unit for combined (remote-base), or individual-unit operations. This packaged unit could thus be used fixed, portable or mobile as desired.

The unique capabilities of a remote-mobile setup with limited access can also produce fascinating results. An amateur can relax at his favorite restaurant or a friend's home while using a hand-held and remote link in his nearby auto for distant communication pleasures. Ultimately, a grocery-shopper in one area of the United States may contact a classroom student in a foreign country, or a lunch-hour operator might contact an afternoon hunting enthusiast hundreds of miles away. The possibilities of simple remote-link arrangements are, indeed, boundless. See Fig. 5-3.

The previously described remote base setup was outlined in Fig. 5-2. Audio output from the 10 FM transceiver is connected to the 2-meter FM transceiver's audio input, and audio output of the 2 meter FM transceiver is fed to the 10 FM transceiver's audio input. Impedance-matching transformers or basic resistance networks similar to that shown are used to match low-impedance speaker outputs to mid- or high-impedance microphone inputs. A basic switching arrangement is used to select either microphone or

Fig. 5-3. Mobile remote-base system in author's auto for weekend fun. Control unit, bottom, selects separate (local) or interconnect (remote) functions. The use of "rubber duckie" antenna restricts the range of 2-meter access. This view is looking back from the driver's position, between the bucket seats. An old coat usually covers the equipment to avoid attracting attention.

remote input for each unit. Carrier-operated-relays, (COR), activated by small voltage changes obtained from the squelch-gate circuitry in each transceiver, keys push-to-talk lines of their respectively connected transceivers. A 12-volt power supply, capable of supplying current for both transceivers and associated circuitry is also included. Two additional items, an on-off tone decoder and a 3-minute timer are also suggested for this setup.

In operation, both transceivers are squelched until activated by signals on the selected 10- or 2-meter frequencies. Assuming incoming signals are received by the 10-meter transceiver, its COR will not place the 2 meter transceiver in transmit status until a discrete tone is first received via the 2-meter receiver. Upon reception of this tone, voltage is applied to both CORs, and remote-base functions are initiated. Signals received via the 10-meter transceiver are thus retransmitted on 2-meters, and vice-versa. If 10-meter received signals continue for 3 minutes without dropping squelch, the timer will return the 2-meter transceiver to receive mode. The remote operator then has the option of tone-disabling the setup, or waiting for the timer to reset. Should the remote operator decide to shut down the system, he merely transmits a discrete tone to the 2-meter transceiver and disables both CORs. A later retransmission of this particular tone will likewise reactivate the system. During periods of non-use or standby activities, all units associated with the remote base are locked in the receive-only mode. Numerous modifications and additions to this basic system are possible. However, such changes will defeat the purpose of this simple-yet-effective setup. There are, for example, no backup provisions in this system. The remote operator must be situated near the system for shutdown provisions during emergencies or malfunctions.

While 10-meter to 2-meter antenna-spacing poses few desensing problems to crossband setups, the antennas should be separated as much as possible to protect sensitive RF-amplifier input circuits. A minimum separation distance of 5 to 10 feet is usually satisfactory for units with reasonably acceptable front-end protection. The carrier-operated-relays and tone decoder may be home constructed, or purchased from amateur equipment manufacturers as desired. While many tone encode/decode capabilities are popular, the amateur desiring maximum security for his setup should consider using odd frequency, subaudible tones, or audible control tones of various frequencies and durations. As previously mentioned, amateurs with absolute control over their system may elect to completely bypass tone control. Remember, however, your reputation (and possibly your license!) may be at stake should unforeseen circumstances create a critical on-the-air situation. Clearing a problem after it happens is also a sign of poor operating and management tactics.

The majority of remote-base setups employ relatively low-power 10-meter FM transceivers operating on an International Direct Frequency such as 29,600 or 29,620 kHz. Since an amateur's existing 2-meter transceiver is usually used for the remote link, it's frequency may be changed as desired. The only provision for this change is ensuring the portable unit used for accessing the remote base also operates on the chosen frequency. Alternatively, two 220-MHz or 440 MHz transceivers may be used in conjunction with the 10-meter FM unit. In fact, any two amateur FM units capable of reliable communications may perform remote capabilities quite feasibly. Amateur ingenuity (while remaining legal) is the keynote to such expansion thoughts.

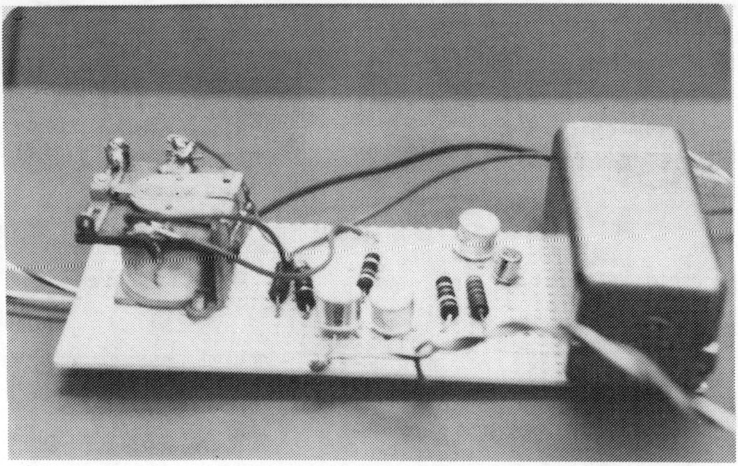

Fig. 5-4. This small circuit board contains the two CORs used in the quick and easy remote-base station. Junk-box parts usually perform well in this flexible circuit.

Assuming the remote base illustrated in Fig. 5-2 is assembled to employ plug-in cables while using a small control box to house tone decoder, CORs, etc., any or all associated equipment may be used separately or together, stored, or moved with minimal effort and forewarning.

The choice of ac power supply or constantly charged battery (which is quite useful during emergencies) is left to the remote base owner/operator.

THE CARRIER-OPERATED-RELAY

The prime element necessary for interconnecting two units to implement remote base or repeater functions is a carrier-operated-relay. Squelch-gate voltage changes acquired from a unit's receiver section activate this switch, which in turn keys the push-to-talk circuit of a separate transmitter. The carrier-operated-relay is usually a 2- or 3-stage device, capable of amplifying low-level voltage variations to a sufficient relay-activation level. Suitable units may be home constructed or purchased from commercial manufacturers, see Fig. 5-4.

A basic carrier-operated-relay might consist of two transistors connected in a Darlington manner, similar to that shown in Fig. 5-5. Squelch-voltage changes applied through this circuit's 56k-ohm resistor control the base of Q1 which, in turn, controls Q2's conduction. When Q2 conducts, the relay activates and ener-

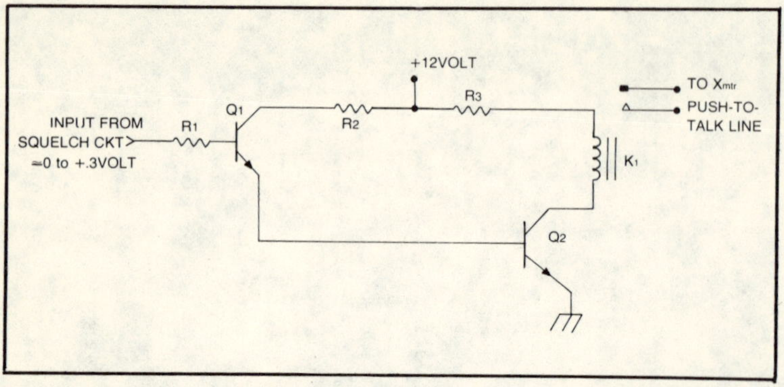

Fig. 5-5. A basic carrier operated relay (COR). R1—56 k-ohm; R2—47 ohm; R3—680 ohm; Q1, Q2—2N697, 2N2222, etc.; K1—low-current, sensitive relay. Approximately 20 k-ohm coil; All resistors are ½-watt, 10% tolerance.

gizes the affected transmitter's push-to-talk line. The 56k-ohm resistor isolates this circuit and prevents loading of the squelch gate. Since Q1's variations are insufficient for controlling the relay, these variations control Q2's base. Q2's large variations then control current flow in the relay's coil.

A slight variation of the previous carrier-operated-relay is shown in Fig. 5-6. This circuit uses an ordinary N-channel field effect transistor for controlling the output transistor. The high input impedance of FETs is quite beneficial in eliminating loading effects on squelch circuits. FETs operate in a manner similar to miniature vacuum tubes: they are voltage-operated devices requiring high-value drain-load resistances, and they are gate-biased in a manner resembling grid-bias methods.

Note that both of the previously described carrier-operated-relays do not include any form of "hold" provisions to keep the transmitter activated after an incoming carrier disappears. While this feature is somewhat inconvenient for single-band repeaters, it is highly desirable for remote base or cross-band setups.

IMPEDANCE-MATCHING TECHNIQUES

A substantial mismatch of impedance usually occurs when the audio output of a receiver is connected to the microphone input of a unit's transmitter. If, for example, a receiver's output impedance is 8 ohms and a transmitter's input impedance is 600 ohms, a 75 to 1 mismatch occurs. If the transmitter's audio input is high impedance, the resultant mismatch is much higher. These mismatches result in excessive distortion, which must be corrected.

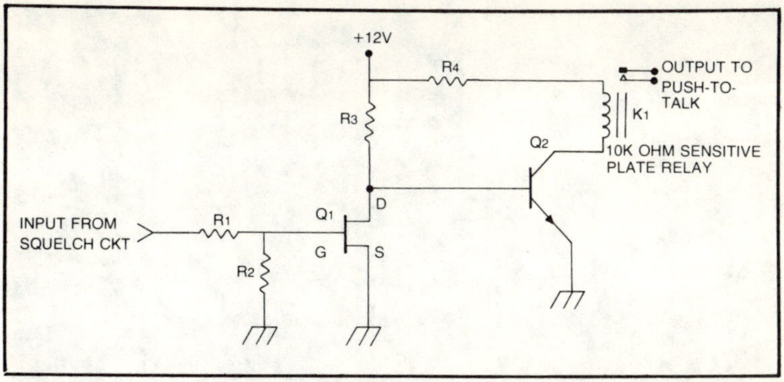

Fig. 5-6. Another COR circuit, using an FET input: R1—1 megohm; R2—1.5 megohm; R3, R4—4700 ohm; Q1—MPF102, or equivalent N-channel FET; Q2—2N697, 2N2222, etc.; All resistors are ½-watt, 10% tolerance.

The most popular method of handling this situation involves placing a matching transformer between receiver and transmitter sections of the particular units employed. The transformer's primary and secondary impedances should match their respective units, and low-wattage transformers are preferred. Since this transformer will act as a load for the receiver's audio output system, its speaker should be switched out of the circuit and audio level reduced when "remote" functions are implemented.

An alternative, and less-expensive means of matching impedances is shown in Fig. 5-7. Two ½ or ¼ watt resistors are used as a high impedance take-off for transmitter audio, and the receiver's speaker is left connected as a load for the audio-output system. The

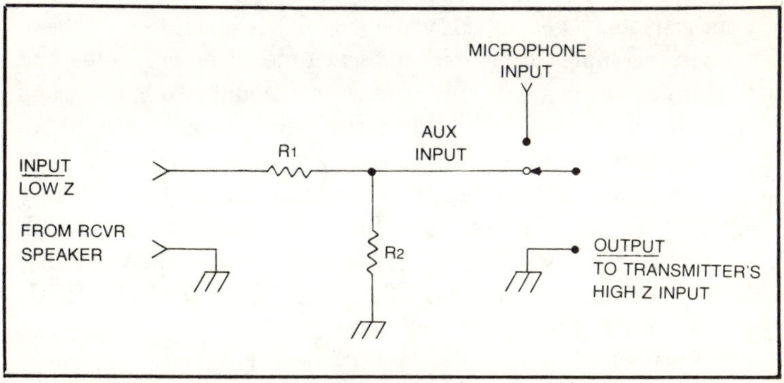

Fig. 5-7. An inexpensive low-to-high input-impedance matching circuit which can be used for interconnecting audio from receiver to transmitter. R1 is 100 k-ohm, R2 is 10 k-ohm, both are ¼ watt, 10% tolerance.

Fig. 5-8. Dave Findley, N6DF, at the receiver site of the WR6BDG 10-meter repeater.

values shown in Fig. 5-7 may be varied to produce desired signal levels, however, their 1-to-10 ratio should be maintained for all medium- to high-input-impedance circuits.

Additional matching techniques are also possible. The only design criteria is merely allowing the associated equipment to see their necessary output/input load requirements while sampling minute amounts of their signals.

THE WR6BDG IN-BAND 10-METER FM REPEATER

Ten-meter repeaters offer many unique possibilities for the FM enthusiasts. The ground - wave and DX capabilities of these machines are highly beneficial both in time of emergencies and during times of pleasure. An extensive amount of planning and operational coordinating is necessary when establishing a 10-meter repeater. Many questions relating to where one begins such an endeavor, how the concept is pursued, and how the necessary equipment is utilized is discussed in the following account, which was supplied by David O. Findley, N6DF, Trustee of WR6BDG, and John E. Portune, WB6ZCT, Technical Director of WR6BDG. See Figs. 5-8 and 5-9.

WR6BDG, Southern California's first regularly operating, in-band 10-meter repeater, has a history dating back to 1976. Credit for its conception goes to John Pike, WA6MSG, who initiated the project with the purchase of a General Electric "Pre-

Prog" rack-mounted, two-unit, repeater/base station. Subsequent discussion with John Portune, WB6ZCT, brought forth plans for a 10-meter repeater with an input of 29.56 MHz and an output on the International Direct Frequency of 29.60 MHz.

As usually happens to all good ham projects, time and cross-town moves by both parties left the equipment in the hands of WB6ZCT at his QTH in the foothills northeast of Los Angeles. After the normal period of delay, the future 10-meter repeater first showed signs of life as a 6-meter to 10-meter crossband device. Combinations included 52.54 in/29.60 out, and 29.56 in/52.525 out. During these preliminary tests, WB6ZCT had an operating 220-MHz repeater at his QTH, and was able to compare the coverage of the higher and lower frequency operations. Ten-meters seemed to show definite advantages at the 1000-foot altitude site, and John was encouraged to experiment further.

In December of 1976, the Southern California Repeater and Remote Base Association (SCRRBA), the 10-meter coordinating body for the Southern California area, coordinated the repeater onto 29.52 in/29.62 out under the call WR6AHW. This call had been in use for some time at the site, as the 220-MHz repeater mentioned above, under the trusteeship of John's brother Frank, N6BW.

Fig. 5-9. John Portune, WB6ZCT, at the transmitter site of the WR6BDG 10-meter repeater. This location is approximately 1 mile from the receiver site. A dedicated telephone line is used for audio linking.

The 52/62 coordination reflected SCRRBA's long-sought-after band plan of 20-kHz channel spacing on the 29.50 to 29.70 portion of 10-meters, where commercial-bandwidth FM repeaters are permitted by the FCC. Much heated conversation was taking place at the time, nationally, by those few devotees to 10-meter repeaters, over alternate 15-kHz plans for the channel spacing, and over the possibilities of SSB repeaters. Fortunately, in the view of this author, the SCRRBA plan has prevailed, and continues to grow toward becoming the accepted national plan.

Also, it was requested by SCRRBA that all coordinated repeaters use continuous-tone, sub-audible squelch (CTS), or as it is more commonly called, "PL" (a Motorola trade mark), to protect their inputs from the intense noise of the band as compared to VHF and higher repeaters. In compliance with this request, WR6BDG is normally only accessible if the received signal has present a CTS tone of 107.2 Hz. All 10-meter repeaters in this area will use the same tone. If an additional SCRRBA plan prevails, each metropolitan area will use a common CTS tone for all its 10-meter repeaters.

Early in 1977, Dave, N6DF, became involved in the activity. Dave lives approximately a mile from John, an ideal separation for good duplex action 10-meters. Although it is technically possible to operate a 10-meter repeater with the transmitter and receiver at the same site, with an input and output separation of only 100 kHz, it is not usually practical. A garage full of water tank sized cavities is usually less desirable than geographically separating the transmitter and receiver.

Initial test operation of the full 10-meter repeater was conducted with a UHF link for the audio and control, composed of Motorola HT-200 Handi-Talkies pressed into this service. The input was on 29.52 MHz and the output usually on 29.60, rather than on the coordinated output of 29.62. The purpose for having it output on 29.60 was to stimulate interest in the local 10-meter simplex community and to solicit coverage reports.

Late in 1977, the UHF link was dropped in favor of a configuration of two, crossband, linked repeaters: WR6AHW, 29.52 in/52.54 out and WR6BDG (licensed to Dave, N6DF) with 52.54 in and 29.62 (usually 29.60) out. Besides being an easy way to link two sites, the dual-repeater concept was to use the 10 meter input and output as a simple, open repeater, reserving the cross-band combinations as a private, fully duplexed autopatch. The autopatch idea never materialized before SCRRBA requested that the 52.54 frequency be vacated for a newly coordinate repeater in San Diego.

WR6AHW/BDG had never actually been coordinated on the frequency.

So at this point, the 29.52 receiver, located at John's house, was linked by a standard, leased, dedicated telephone line to the 29.62 transmitter at Dave's house. The repeater took up its current call WR6BDG.

We might digress a moment to discuss the Southern California location of the repeater. Sierra Madre considers itself a "foothill village", somewhat out of the mainstream, although only 17 miles northeast of downtown Los Angeles. It is nestled against the slopes of the San Gabriel Mountains. Dave's QTH is at an elevation of 700 feet (the Pacific Ocean is about 40 miles distance) and John's QTH is at 1000 feet, about one mile distance. Above, this, the elevation rises abruptly to 5710 feet high Mount Wilson, sporting the famous observatory and a pincushion of Los Angeles' major television-station antennas.

Since February, 1978, WR6BDG has been in full service with few interruptions. In May, 1978, the receiver acquired "new ears" with the addition of a Hamtronics pre-amplifier. The two-frequency (60/62) relay was abandoned in late September, 1978. This perhaps marks the end of the testing stage and WR6BDG was "on its own" on its approved channel of 29.62 MHz out.

In November of 1978, a simple control device was added to permit remote defeat of the CTS for DX stations. This has proved to be a real asset in stimulating activity on the band for repeaters.

In December, the transmitting antenna at Dave's house met its demise in a wind storm. During the repair session the transmitting and receiving sites were switched as an experiment. The results were better local coverage with similar DX performance. Some further experimentation is anticipated.

Philosophy Of WR6BDG

"Southern California is "blessed" with wall-to-wall repeaters and remote bases on its many mountain tops. Most are in the 2-meter and 70-cm spectrums, with a growing community on 200 MHz and 6 meters. High-quality, wide-range local coverage is available on a dozen or more of these systems.

"We wanted to see if good coverage from a lower altitude would be made available by the use of 10 meters. Many months of comparison clearly demonstrated that equal power and receiver performance favored the 10-meter repeater over the 220-MHz repeater at John's QTH. But, the coverage did not compete with

Fig. 5-10. A Radio Shack antenna is used at the receiving end of WR6BDG. The dipole just below the ground plane is for 70 cm.

the real "mountain top" installations. So, although WR6BDG does have a good usable range of 25 miles or so, it has not attracted a local usership of much more than the original core of builders. However, it has recently begun to attract considerable DX usage.

"That is why we now consider its future as principally a DX machine, to be used more by non-Californians than locals. It is really quite novel to listen to two hams talking to each other virtually 'across town' through a repeater 2000 miles away. The sunspot maximum of the later 70s and early 80s will undoubtedly provide some truly unique QSOs through the system.

So we are quite content to leave the system at lower altitude and 'let the DX roll.' Here, low noise levels are superior to higher altitudes, and WR6BDG can out-hear mountain top systems in respect to DX. See Fig. 5-10.

Technical Aspects Of WR6BDG

From the electronic vantage point, WR6BDG is a marvel of antique state-of the art. The bright glow of 807 filaments and the clank of large relays continuously emanate from its garage-corner location. And, except for the anomaly of a solid-state identifier, its inner workings are true to the vintage of its 1950s model, 40-MHz PreProg components. But, in true ham fashion, it has "worked out well at this QTH" and been a faithful warmer of the ionosphere.

Fig. 5-11. The receiver section of the WR6BDG 10-meter repeater.

Retuning the large phenolic coils from 40 to 29 MHz was an easy chore, and the interconnection of the transmitter and receiver was equally simple—a leased telephone pair. Two suitable vintage oil-filled 10 μF, capacitors feed and receive audio from the line. The receiver came with separate speaker and 600-ohm line outputs, already correct for repeater service. The WR6BDG receiver is shown in Fig. 5-11.

At the transmitter, audio is simply inserted in the microphone amplifier at a high-level stage through a potentiometer and a high-value series resistor. The original station microphone thereby remains as a local microphone.

Both a carrier operated (squelch) relay and a CTS relay are in use. Their contacts are in series to give maximum immunity from various band "gunge." Both must kick in to put the transmitter on the air. However, the CTS contacts can be shorted locally or remotely to defeat the CTS protection as mentioned above. The CTS frequency is 107.2 Hz, and must be continuously present on the received signal with a deviation of about 700 Hz. This is decoded at the receive site by a standard GE Progress Line reed decoder, fed directly by the receiver's discriminator. See Fig. 5-12.

At present, full CTS protection is used during the week. On weekends, the CTS is usually defeated. Continuous COR-only operation is out of the question on this band in all but the most unusually quiet locations. Overlooking the great industrial basin of Los Angeles, even at a low altitude, the RF noise is horrible. Older-type RF plastic-bag-sealing machines are the principle nightmare—sweeping their raspy way across the 10-meter band with such monotonous regularity as to make the hapless COR-only 10-meter repeater key up like the proverbial popcorn machine.

Keying of the transmitter is accomplished over the same telephone pair by the simple expedient of applying 18 volts DC through two small audio chokes. Some modest audio shaping is also performed by a small perfboard® bristling with resistors, capacitors and 88-mH toroids. The line is not flat, but gives nearly ideal 300- to 3000-Hz response when combined with some additional frequency tailoring in various of the other audio stages.

The CTS remote defeat device, called on more pompous days the control system, is a true-blue 1950s vintage stepping relay pulsed by the COR and the CTS relays. To make it work, bring up a carrier with CTS on it. Then briefly drop the CTS once to activate

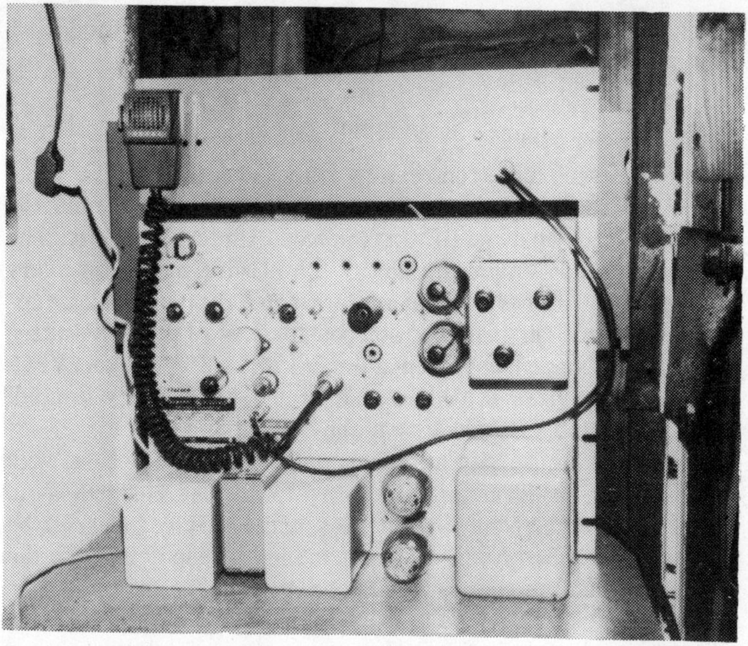

Fig. 5-12. Transmitter portion of the WR6BDG repeater. A garage corner provides housing for this elderly example of classic electronic technology.

CTS and twice to defeat it. A touch-tone decoder was deemed out of character (actually, too expensive.)

WR6BDG's ubiquitous presence on 10-meters is maintained by a simple TTL identifier keyed every 15 minutes and once every three minutes under COR control. Even though you may not have CTS on your signal you will be able to key up the identifier and talk while it is playing. This may help you raise a control operator to turn off the CTS.

For a busy local repeater, such a frequency ID would truly "harass the troops." But in the new world of DX repeaters, the endless bubbling of this little black box of silicone slabs has, and will continue, to contribute to the education of the international 10-meter FM DX community. Without it, the presence of a repeater would go unnoticed as the newcomer glides by on his VFO. Also, hopefully, it will dissuade the frequency SSB invader who for some reason finds it hard to find a clear spot (except on the input of the repeater) anywhere else on the hundreds of possible simplex SSB frequencies on 10-meters. But, we give them the benefit of the doubt of not being "clued in" on the activity, and the machinery which loves SSB signals, clattering away on that minute window of 10-meters where FM repeaters are permitted.

On-The-Air-Experiences Of WR6BDG

Two forces have contributed to the increased utilization of WR6BGD recently—the peaking of the sunspot cycle and the introduction of Japanese HF transceivers with FM capability. The international FM simplex frequency, 29.60 MHz, has become very busy recently but has been largely populated by retuned commercial FM transceivers using crystal control. Few of these permitted the user to happen upon repeaters off of 29.60. But the VFO-controlled Japanese transceivers have not been thusly limited. Also, converted CB radios have played a small part.

Most of the activity has been confined to U.S. stations with virtually all call districts being heard from. The best DX as of mid-1979 has been New Zealand. Most of the activity appears to be stations in the South and Midwest talking to each other via the repeater when the short-skip direct path is not open.

One humorous incident bears repeating. Recently an active 10-meter FM group in the St. Louis area discovered the repeater and began working stations all over the South and Midwest. In the process, they would hear a dog barking in the conversation. John, WB6ZCT, had to come in and tell them that the local microphone is live during repeat and you may occasionally hear his dogs, or other local reverberations. So the mystery dog was identified.

Future Plans Of WR6BDG

"The need for continuous-tone squelch to keep out unwanted noise may become a deterrent to increased activity, certainly to the casual newcomer. Therefore, we are considering some form of "whistle-on" or "touch tone" feature for the future. This would permit 3 minutes of operation without CTS.

"With regard to operating technique, when the repeater is not busy, we encourage long calls or calls in which the facts of the repeater are given. Continuous monitoring of any channel other than 29.60 is still rare, so only a longer call or an on-going conversation is likely to catch a knob twiddler tuning off of 29.60. So, don't hesitate to rag chew and give your call and the repeater's frequencies several times when you call."

Chapter 6
Antennas for 10 FM

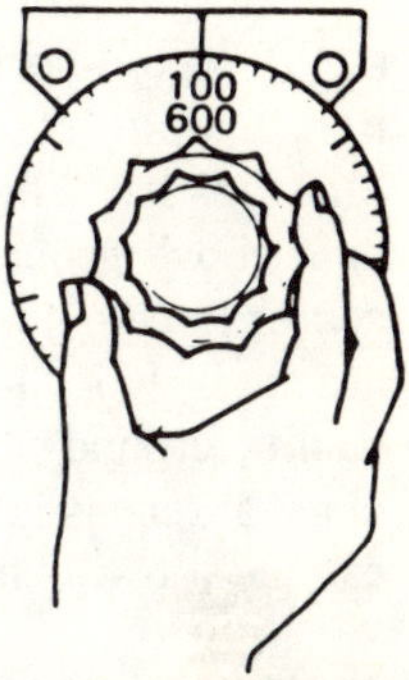

All radio amateurs realize that an efficient antenna system is the key factor to successful long distance communication. Numerous examples acquired through years of actual on-the-air experiences continue to prove the accuracy of this statement. Conventional amateur setups with superior antenna systems consistently outperform more elaborate, high power stations using "compromise" antennas. Additionally, the weak-signal reception capabilities of an antenna system is directly proportional to its overall gain factor. A less expensive means of increasing the amateur station DXability is, indeed, difficult to visualize. While lavish skywires may be beyond the realms of feasibility for low-frequency amateur-band operations, their use on 10-meters is often quite acceptable. The reasoning behind this statement is really quite simple. The size of a shorter-wave -length 10-meter antenna is approximately one-half its 20-meter counterpart. While the nature of amateur radio's 10-meter band supports unlimited enjoyment when using basic antennas of unit gain, larger arrays can substantially extend one's horizons during prime band openings. The foremost consideration of this aspect is cost.

Ten-meter FM has one other parameter in its favor: antenna height above sea level. While portable operations from mountain peaks on the 80- or 40-meter amateur bands are mildly advantageous, 10-meter mountain-topping activities produce fantastic results. Now let's relate the previously mentioned factors to 10-

meter FM amateur setups and see what antenna systems might prove worthwhile. See Fig. 6-1.

Vertical antennas, one-quarter wavelength or larger, appear to perform better on 10-meters than they do on 20-or 15-meters. This is probably due to the fact that vital one-half wavelength "radiating space" is easier to obtain on the higher frequency band. Verticals will not operate efficiently, however, unless they have an unobstructed-horizon view in the desired directions of communication, and an effective ground system is used. It should also be realized that single-element verticals mounted on a metal mast high above ground are not true vertical antennas but rather off-center fed dipoles of limited performance. Ground-plane antennas are not included in this category, since their radials establish a high quality ground plane at any height. The performance of ground-plane antennas is usually between that of a ground-mounted vertical and a tower mounted beam or quad.

BEAMS

Tri-band beam antennas perform quite well on 10 FM. As mentioned in a previous chapter, a simple antenna tuner may be necessary for reducing the SWR at the 29-MHz end of 10-meters. Many tri-band antennas produce comprising results on bands other than 10-meters, and unsuspecting amateurs may be pleasantly surprised with the performance of their outdoor treasure in this respect. A narrower beam width and greater front-to-side rejection will also be noticed on the "high end" of 10-meters. Optimum results are thus obtained by using omni-directional antennas for determining directions of band openings, then switching to a beam antenna for consequent operations.

WIRES

Longwire antennas, matched with any of the popular antenna tuners, also produce very good results on 10 FM. These antennas are an excellent choice for portable or vacation activities when very little time is available for erecting conventional radiators. A very long wire may be used to fly a kite, or shoot over treetops with a bow and arrow, and connected to an antenna tuner. The tuner can be adjusted for a 1-to-1 SWR within a couple of minutes time, and high performance 10 FM activities may commence. While longwire antennas of several wavelengths are difficult to erect for the 20- or 15-meter bands, 10-meter longwires are easier to put up and tune to resonance.

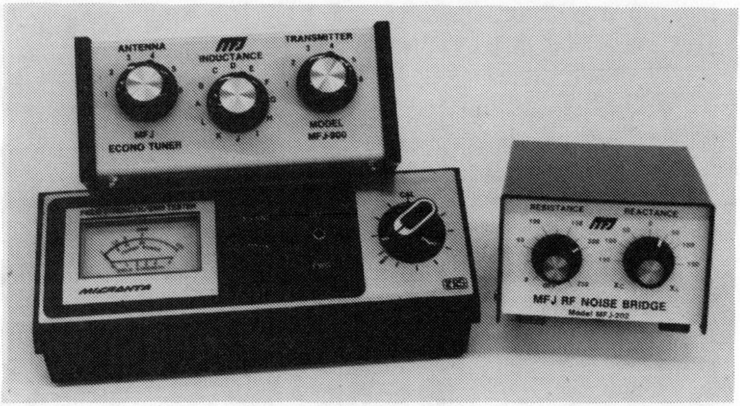

Fig. 6-1. The basic test instruments and operating aids for 10-meter FM are similar to those used on any Amateur high-frequency band. Shown are an antenna tuner or matcher, an SWR meter, and a noise bridge to aid in adjusting antennas.

While Citizen Band antennas trimmed for minimum SWR at 29 MHz make ideal 10-meter FM antennas, unmodified CB antennas also produce good results when used with an antenna tuner. Low power operations on 10 FM also permit the use of small, inexpensive antenna tuners. In fact, many "CB tuners" will perform this function very well.

While the reader is probably beginning to visualize his own special antenna for 10 FM, some specific ideas will now be described to further compliment those thoughts.

THE HELICAL-ELEMENT MINI-BEAM

One of the more popular CB mobile antennas presently being marketed is the helical-wound "wonder shaft" mobile whip. These whips can be tuned to 29 MHz by carefully trimming their length in one-quarter-inch steps while monitoring their SWR and resonant frequency (Fig. 6-2). An antenna noise bridge and CW receiver tuned to approximately 29,650 kHz may also be used for this operation.

Assuming the 10 FMer either acquires four "wonder shaft" whips from ex-CB'ers, or winds his own equivalent elements using copper wire and fiberglass fishing rods, he has the basic elements to construct an efficient two-element beam antenna. Two whips are tuned to resonate at approximately 29,650 kHz. These items will compromise the beam's reflector. Next, the elements are affixed to 2 by 2 wooden supports which are bolted to an associated six-foot

Fig. 6-2. A sabre-saw can be used to trim a glass-fiber helical whip for 10-meter FM. The whip was originally used on the 27-MHz CB band. Trimming should be done in very small increments to avoid overcutting.

wooden beam. A floor flange connected to this boom then supports the complete array. A W2AU or Amidion balun, which matches 50-ohm input to output impedances, may be connected to the driven element. Specific construction/layout methods vary according to individual desires, thus precise mechanical details have been omitted from this general description (Fig. 6-3).

Amateurs desiring to construct a full-size, two-element beam for 10 FM may salvage elements and parts from a damaged or discarded CB beam, and follow the previously described ideas. Assuming full-length elements are employed, each side of the driven element should be approximately 94 inches and each side of the reflector should be approximately 98 inches. After construction and installation, a final antenna adjustment for maximum forward gain may be performed with the aid of a simple field-strength meter.

THE MINI CURTAIN

A relatively simple wire antenna capable of outstanding performance on 10-meters is the Bruce Array. The typical gain of this curtain-type array is 5 dB, and its low angle of radiation is highly beneficial for long-distance communications. The antenna is ideal for portable or fixed-station use; it can be rolled into a small bundle for easy transporation, and it may be constructed from surplus lengths of wire to reduce cost. A sketch of the Bruce Array is shown in Fig. 6-4.

Each section of the Bruce Array employs a full wavelength of wire, yet the antenna's overall length is maintained at one full wavelength. The "capture area" afforded by this setup is thus enhanced substantially. The vertical phasing sections, labeled A in Fig. 6-4, may be constructed from conventional 300-ohm TV twin lead, while horizontal sections, labeled B, and end sections, labeled C in Fig. 6-4, may be constructed from conventional bare or insulated, stranded wire. The two vertical phasing sections are twisted one-half turn while the middle section is not twisted. A 50-ohm unbalanced input to 300-ohm balanced output balun transformer is connected to the untwisted center section's exact middle. Suitable baluns are manufactured by companies such as Amidion. High-quality, 50-ohm coaxial cable may then be connected between the 10 FM unit and the antenna.

Once the antenna is ground-level constructed, and all connections are secured, the array may be raised into position. Locate a clear area with tree limbs capable of supporting the array. Next, using a weighted object for throwing, or a bow and arrow, pass support ropes over the selected limbs and allow excess rope to fall back to ground on either side of the limbs. Finally, connect the array to the support ropes and raise the antenna into position.

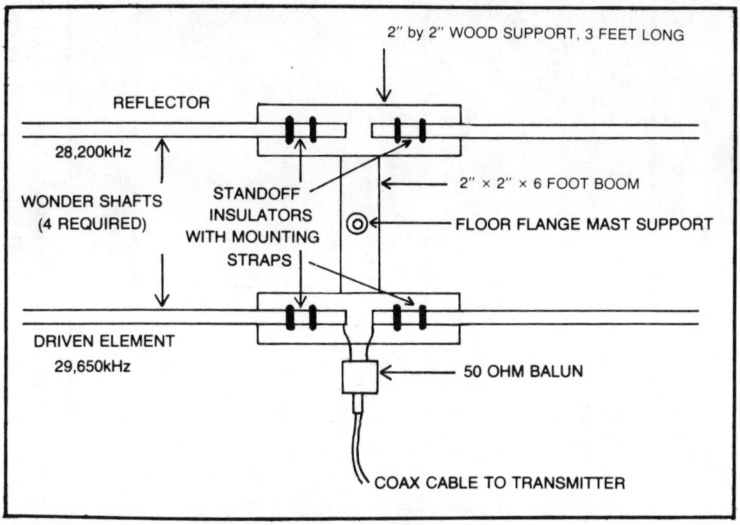

Fig. 6-3. Mechanical details of a helical element minibeam for 10-meter FM use. The boom may be wood, or sections of shower rods.

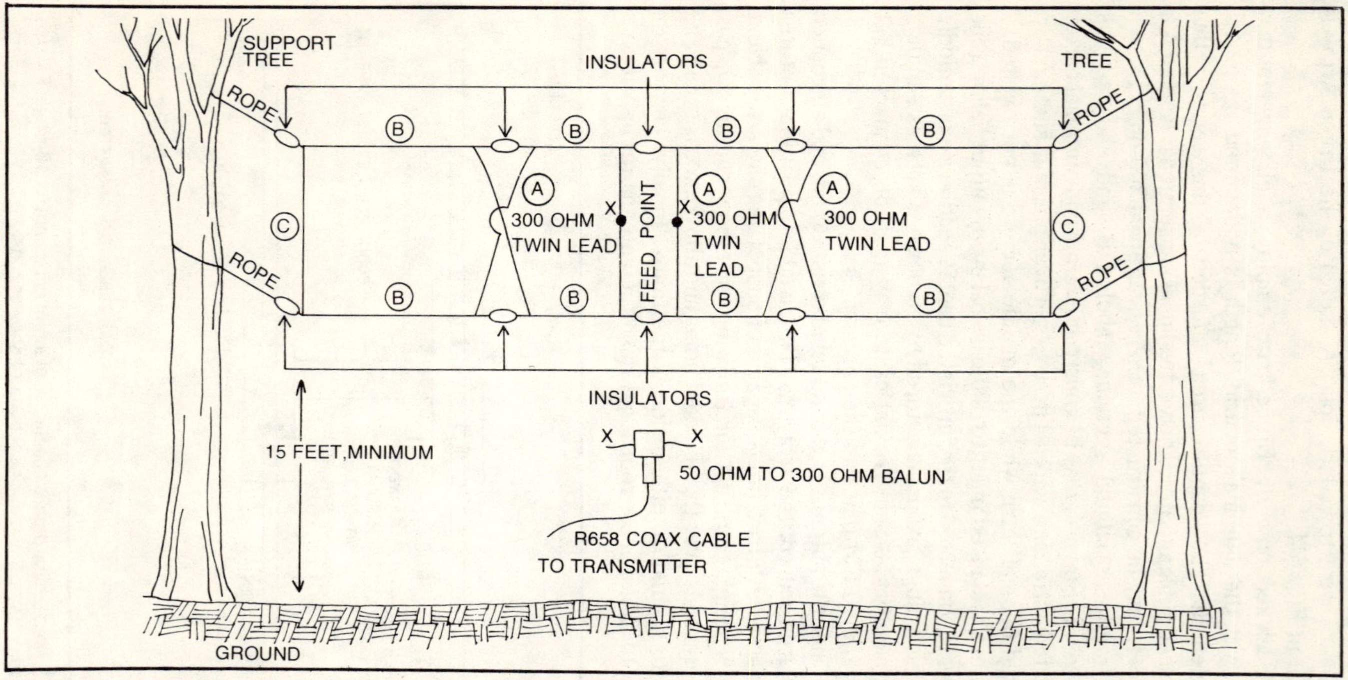

Fig. 6-4. The mini-curtain antenna. Lengths "A" are ¼-wavelength, or 8.7 feet of 300-ohm ribbon; Lengths "B" and "C" are 8.7 feet of stranded wire. The bottom of the array should be at least 15 feet above the ground.

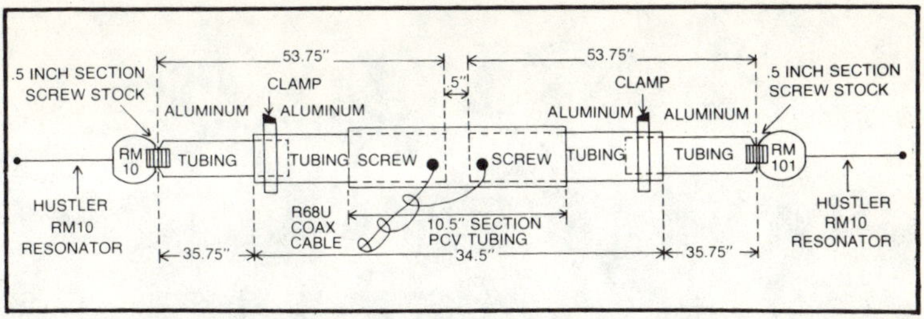

Fig. 6-5. The knapsack dipole. All sections of tubing and the RM-10 resonators may be secured in a compact package and wrapped with tape for easy transportation.

The dimensions given in Fig. 6-4 will allow coverage of approximately 28,600 to 29,700 kHz with an SWR of less than 1.7 to 1. If a lower SWR is desired, each horizontal section may be trimmed in one-half-inch increments. These adjustments should begin with outer sections and progress toward the center as necessary.

THE KNAPSACK DIPOLE

There are many times when the 10 FM entusiast would like to enjoy operations from a spontaneous location or weekend vacation cottage, but he hasn't the leeway to erect a wire antenna. Prospects of using the auto's mobile antenna are usually discouraged

Fig. 6-6. K4TWJ setting up the knapsack dipole for signal comparisons and tests during a weekend band opening. Note the use of Hustler 10-meter resonator taken from the mobile antenna on car.

Fig. 6-7. Details of center connections and clamping arrangement used on the knapsack dipole.

due to nearby obstructions, lack of altitude, etc. One convenient solution to this dilemma involves using a knock-down rotary dipole with mobile resonators at each end of the sectioned aluminum elements. The popular New Tronics Hustler mobile resonators are perfectly suited for such applications. Since the mast sections used with all New Tronics mobile resonators are 54 inches in length, each side of the dipole must attain this length. Likewise, threaded rods to accommodate the resonators can be inserted into the tubing's outer ends. This will allow a resonator removed from the mobile antenna's mast (plus a spare, borrowed from a friend) to mate with the rotary dipole and complete the antenna. A further description of this antenna is shown in Fig. 6-5. The completed dipole may be clamped to predrilled wooden or aluminum mast, and mounted at a vantage point during operations. See Figs. 6-6 and 6-7.

Pursuing the knapsack-dipole concept a step further, a second element and two more resonators could be added to form a portable beam of outstanding capability. Items such as boom, reflector element mounts, boom to mast plate, etc., could be salvaged from discarded CB antennas. All element sections can be color coded, and length-adjustments marked for quick in-field assembly. Parasitic elements should be tuned for resonance at their appropriate 10 meter frequencies. An antenna noise bridge and CW receiver are

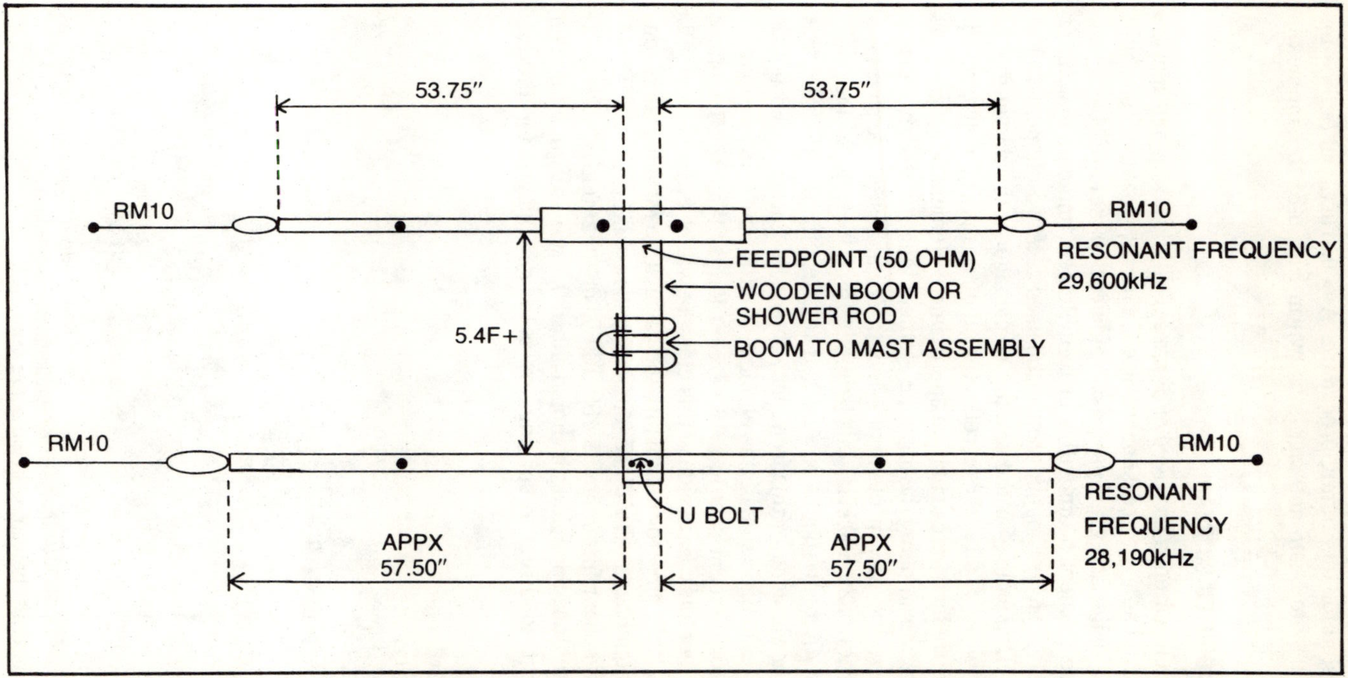

Fig. 6-8. Construction details for a portable beam using Hustler resonators and short lengths of tubing for compact storage. An antenna noise bridge may be used for initial antenna tuneup.

suggested for this tuning procedure. A constructional outline for this portable beam antenna is shown in Fig. 6-8; the antenna is shown in Fig. 6-9.

THE MODIFIED CB BEAM

Quite often, a damaged or discarded CB beam antenna may be purchased at a very reasonable price. The antenna may be rebuilt into a smaller array and used as an efficient 10-meter FM antenna. A three-element beam with a broken element and beta match, for example, can be chopped in size and used as a two-element 10 FM beam. Depending on the amount of damage to the parasitic element, either a director or reflector element may be constructed. The procedure for calculating approximate element lengths are as follows. First, using the formula: element length = 468/frequency in MHz, the driven element's length is calculated. Assuming a frequency of 29,600 kHz, this length is 15.81 feet. Each side of the driven element should be one half this length, or 7.905 feet. A director element for this frequency should be 4 precent shorter, or 15.17 feet. Each side of this element will thus be 7.58 feet long. A reflector element should be 5 percent longer than 15.81 feet, or 16.60 feet long. Each side of this element will be 8.3 feet in length. Reinserting director length into the calculation formula, its resonant frequency is 468/15.17, or 30,850 kHz. Likewise, the reflector's resonant frequency is 468/16.60, or 28,200 kHz.

Antenna boom length should afford 0.1 wavelength for driven element to director spacing, or 0.15 wavelength for driven-

Fig. 6-9. The knapsack dipole becomes a high-performance beam with the addition of a reflector element. All sections dismantle into a small package which can easily be carried in the auto's trunk.

element to reflector spacing. These calculated length are 0.1 × 2(15.81), or 3.16 feet for the director and 0.15 × 2(15.81), or 4.74 feet for the reflector. The forward gain of a reflector/driven element array is usually 5.5 dB, while a director/driven element array usually produces 4.2 dB.

Once the antenna is constructed and its elements are adjusted for proper lengths, all resonant frequencies, and the antenna impedance, may be checked with the aid of an antenna noise bridge and general-coverage receiver. A final forward-gain touch-up may be performed, if desired, while using a field-strength meter placed approximately one wavelength from the antenna.

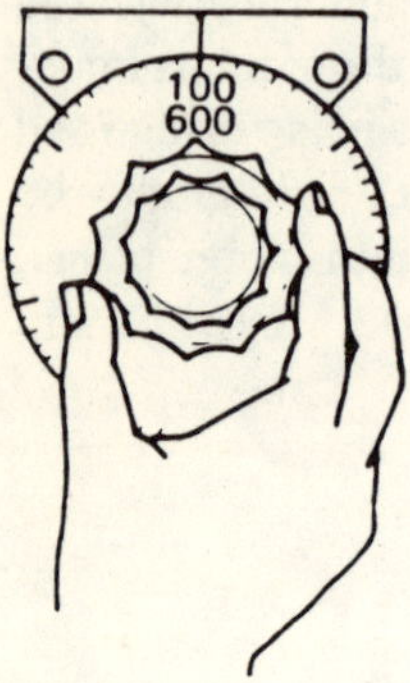

Chapter 7
FM Theory

Newcomers to amateur radio may wonder why FM is so extremely popular, how it gained widespread acceptance, and what are its basic differences compared to AM or SSB. This chapter will attempt to clarify those vague areas with a simple, straightforward, and semi-technical explanation.

Contrary to popular belief, Frequency Modulation has existed since the mid 1920s. Since FM transmission required a large bandwidth, however, it didn't acquire general acceptance until several years later. During 1936, Major E. H. Armstrong presented a disertation to the Institute of Radio Engineers concerning the use of his newly developed wideband FM concept which used the recently pioneered higher radio frequency bands, VHF and UHF. The series of events and innovations that followed this "first step forward" ultimately resulted in a large-scale acceptance of FM on these higher frequency bands. One evolution during this time was the development of a narrowband FM system. While wideband FM continues to enjoy a prime spot in commercial high-fidelity broadcasting, narrowband FM has become the accepted mode for VHF and UHF communication systems, which range from police and public-service operations to amateur radio applications.

Narrow-band frequency modulation occupies a wider bandwidth than a single-sideband signal, thus its use on amateur frequencies below 28 MHz is quite illogical. Since "skip" conditions are less frequent on VHF bands, numerous services may use

FM with minimal cross-interference problems. Likewise, the noise immunity afforded by FM communications provides an efficient and professional capability which is unobtainable with amplitude modulation or single sideband systems. It should be realized this noise immunity is the result of amplitude-limiting stages included in the FM receiver's I-F strip, not in the use of FM transmitters proper. All communications modes are affected by extraneous noises, however, FM conveys intelligence in the form of frequency deviations from an established center frequency rather than amplitude variations of a specific-frequency signal. This relationship is illustrated in Fig. 7-1. Another special benefit of amateur-radio employed narrow band FM is its simplicity of modulator design. Large modulation transformers and/or complex modulation systems may be replaced with simple and inexpensive varactor circuits which merely shift the transmitter's carrier frequency at an audio rate. Since this technique employs a carrier which doesn't necessarily vary in amplitude with intelligence, highly efficient class-C amplifier circuits may be used throughout RF stages.

TERMS ASSOCIATED WITH FREQUENCY MODULATION

Several exclusive terms are used to describe the various parameters associated with FM. Since the FM newcomer may not be fully familiar with these terms, a condensed technical description will be presented at this time.

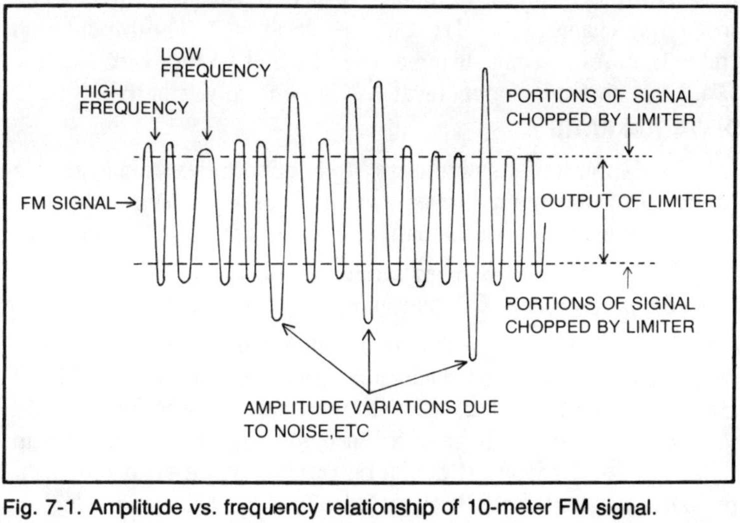

Fig. 7-1. Amplitude vs. frequency relationship of 10-meter FM signal.

An unmodulated FM carrier is merely a continuous-wave carrier, its frequency being designated the carrier frequency, or center frequency in FM terms. As modulation is applied to this carrier, its frequency swings, or deviates above and below its center frequency ata a rate approximately equal to that particular modulating frequency. In other words, the *rate of swing* (or rate of *deviation*) is proportional to the *frequency* of the modulating audio signal. The *amount* of carrier swing (amount of carrier deviation) is proportional to the *amplitude* of the modulating signal. This means that high-amplitude modulating signals will deviate an FM carrier over a wider frequency range than will low-amplitude signals. Notice that the amount of deviation (modulation amplitude) and rate of deviation (modulating frequencies) are independent variables. A further understanding of FM techniques may be acquired by the introduction of two additional terms: Modulation Index and Deviation Ratio.

MODULATION INDEX

Modulation Index is defined as the ratio of frequency-modulated-carrier swing to the audio modulating frequency. Mathematically this is described as follows:

Instantaneous frequency swing =
(modulation index) X (frequency of audio modulating frequency)

As an example, assume an audio modulating frequency of 2000 Hz produces an instantaneous frequency swing of 2 kHz. 2000 ÷ 2000 = 1. The Modulation Index is 1. Next, assume an audio modulating frequency of 2000 Hz produces an instantaneous frequency change of 4 kHz. 4000 ÷ 2000 = 2. The Modulation Index is 2. As you can surmise, the Modulation Index for Narrow Band FM amateur communications is generally less than 3 to 1.

DEVIATION RATIO

Deviation Ratio is similar to Modulation Index, however it is the ratio of maximum carrier swing to the maximum audio frequency transmitted. Mathematically this is described as follows:

Maximum carrier swing =
(deviation ratio) X (maximum audio frequency utilized)

As an example, assume that a maximum audio modulating frequency of 3100 Hz produces a maximum carrier swing of 7200 Hz. 7200 ÷ 3100 = 2.322. (Remember we are considering absolute maximum values in these calculations). The deviation ratio is 2.322. In some respects, deviation ratios may be visualized as a phenomena similar to PEP rating of Single Sideband. Due to individual voice characteristics, this ratio can vary between different amateurs and their equipment. This consideration is, however, quite different from SSB similarities, in the resepct that no RF power fluctuations are generated in FM systems. When penetrating voices are impressed on FM systems, the result is merely an instantaneous frequency over-deviation. This introduces the possible consequence of distortion of the transmitted signal. A possible misconception may exist at this point, thus the following aspect must also be considered: It is possible for 10-meter FM signals traveling long distances to inherit a miniscule amount of distortion, regardless of the originally transmitted signal quality. This possible distortion results from multipath propagation producing phase shifts of the FM signal. Such conditions are not unknown to avid DXers and slow-scan TV operators. FM DXers, however, may be puzzled by this phenomena during the first few times it is actually experienced.

The sidebands produced by an FM signal are quite different from those produced by AM or SSB systems. As an example, assume that an AM signal is modulated by ordinary speech. This means the steady-frequency carrier is varied in amplitude at an audio rate. As a result, an upper and lower sideband are produced; each containing energy proportional to the modulating signal. Now let's assume that an FM signal is modulated by ordinary speech and analyze the resultant difference. During frequency modulation transmissions, the carrier's amplitude remains constant while its instantaneous frequency fluctuates quite rapidly to follow the modulating signal. These extremely fast fluctuations move the carrier above and below center frequency, producing numerous new frequencies, or sidebands. Each of these sidebands contain different amounts of RF energy. While many FM sidebands containing large amounts of energy are produced, many additional less significant sidebands containing smaller quantities of energy are also produced. Most of these sidebands, however, are required to reproduce the originally transmitted signal. If we could see a hypothetical FM signal, the previous effect could be visualized. Without modulation, the steady carrier would rest at one particular frequency and would not vary. With modulation, rapid dulations of the carrier would fill the bandpass in a "strobe effect" of signals. As we viewed this spectrum, some of the signals would be more defined than others. Different amplitudes might also be apparent, however these amplitudes would not have any actual effect on the reproduced signal. If an extremely selective field of view could then be moved across our hypothetical signal, we would see each frequency in the bandpass changing energy content according to the modulating signal. Thus we see that intelligence conveyed by an FM signal is contained in its multiple frequencies which are produced above and below the carrier's center frequency during deviations.

CONCEPT OF NARROW-BAND FM

Narrow Band Frequency Modulation has become a popular mode of amateur communications for several reasons. The most important of these reasons will now be discussed. Small amounts of frequency deviation are easily accomplished through the use of inexpensive varactors. Varactors, or varicaps, are voltage-variable capacitors: their capacitance changes ever-so-slightly according to the DC voltage impressed across their terminals. A varicap connected to the oscillator circuit of an inexpensive CW transmitter will thus produce an FM signal. Since intellignece is conveyed in frequency variations of the resultant carrier, rather than in a modulation envelope, inexpensive and highly efficient Class-C amplifier stages may be utilized.

Detectors used in narrow band FM differ from AM counterparts in the respect their output variations follow frequency changes rather than amplitude variations. This is usually accomplished by using the equivalent of two detectors circuits: one resonant above the I-F center frequency and one resonant below the I-Fs center frequency. This permits the resultant output to vary in amplitude as the incoming signal swings toward and away from the resonant frequency of each tuned circuit.

Since the bandpass of NBFM is comparable to that of conventional AM, special low-*Q* circuits, or bandpass swamping techniques, are seldom required.

MEASURING NBFM PARAMETERS

Each mode of amateur communications has its own set of parameters which are measured in various ways, and FM operations are no exception to this rule.

One of the simplest, yet most important, FM measurements concerns determining RF output power. Since FM transmissions produce a CW carrier, one simply keys the microphone's push-to-talk line and reads this value on a wattmeter. The particular unit's RF stages may be peaked for maximum output at this time, if desired. It should be understood that service instructions are to be diligently followed during this procedure.

Another vitally important measurement concerns assuring FM transmissions are exactly on the desired frequency of operation. While very slight off-frequency operations may be acceptable for local or strong-signal communications, amateurs striving to enjoy maximum distance capabilities from their equipment must maintain exact-frequency relations. Frequency counters, or auxiliary receivers, accurately calibrated against WWV may be used for 10-meter FM checks. The procedure simply involves connecting a dummy load to the 10-meter transmitter, and sampling a small amount of its output under no-modulation conditions. Several of the new digital-readout, amateur high-frequency transceivers are perfectly suited for such measurements. Care must be exercised to protect the receiver's RF section of these units; disconnecting the antenna, and shorting receiver input terminals usually affords sufficient attenuation of output from nearby 10 FM units under test.

An accurate determination of FM receiver performance may be derived by using the SINAD method of sensitivity measurement. The term SINAD relates to signal, noise, and distortion measurements of a receiver. To perform this test, an RF signal generator which is modulated by a 1000-Hz tone at 3-kHz deviation is connected to the receiver's antenna input terminals. A vacuum-tube volt-meter is connected to the receiver's speaker, and the squelch control is turned to the unsquelched position. An adjustable audio filter having an attenuation of at least 40 db at 1000 Hz and 0 db below 900 Hz and above 1100 Hz is also connected to the receiver's audio output. The volume control is set near full output.

Initially, the signal generator is set for a high RF output level and tuned until the receiver's discriminator meter reads zero, or a maximum reading is obtained on the speaker-connected vacuum tube volt-meter. Next, the signal generator's output is reduced to 0.4 microvolts and the resultant audio-output level is noted. The 1000-Hz filter is switched into the circuit, and the audio-output level is again noted. If the difference is 16 dB, the 0.4 microvolt generator level is the receiver's sensitivity. If the difference is more or less than 16 dB, the generator's output must be varied until the receiver's unfiltered audio output is four times greater than the noise-plus-distortion output with the 1000-Hz filter removing the audio modulation. The resultant generator output level will be the receiver's SINAD sensitivity rating.

Mathematically stated, the previously described SINAD measurement may be illustrated as follows:

$$\frac{\text{Signal} + \text{Noise} + \text{Distortion}}{\text{Noise} + \text{Distortion}}$$

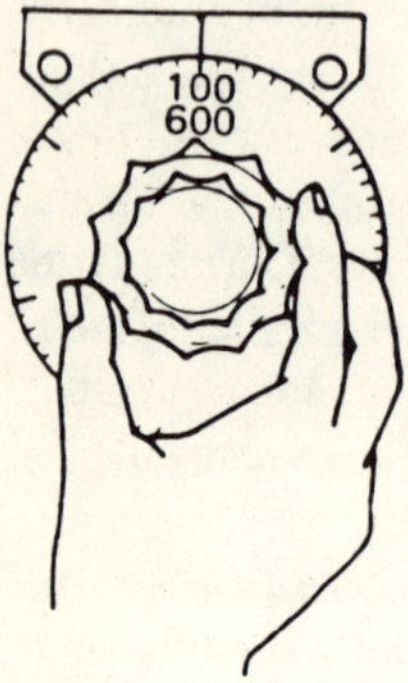

Chapter 8 Medium-Scan TV and 10 FM

During the early part of 1978, a small group of slow-scan TV enthusiasts began pioneering efforts for a new and unique form of visual communications. The dreams, or visions, of this group was the capability of exchanging live motion pictures on an intercontinental basis. A vision of this nature was not a whimsical venture, since the ultimate results would provide a concept heretofore deemed impossible or impractical. Likewise, the initiation and involvement in a project of this magnitude doesn't necessarily assure absolute success. Similar projects sponsored by commercial sources have fallen short of achieving comparable results. There are two possible reasons for this situation. First, the home experimenter can enjoy the liberty of pursuing goals which large corporations abandon due to political tactics, financial distributions, or time allocations. The determination to give "impossible dreams" one more enthusiastic try has proven its worth in true amateur spirits many times. Second, the advent of digital electronics has opened the door to many techniques which could not be achieved at the time original criteria were established.

Millions of tv sets in use by the general public can't simply be outmoded for radical parameter changes. Modern amateur-radio systems, however, are not bound to such outdated restrictions. A brief investigation of present television techniques will further clarify this point of discussion and provide a basis for medium-scan TV applications.

Due to the large amount of information content and high scanning rates used in conventional fast-scan Television, a bandwidth in excess of 3 MHz is required for video transmission. The RF bandwidth required to transmit any form of intelligence is directly proportional to the quanity of information contained in that signal. Voice signals require a wider bandwidth than CW signals because the conveyed range of voice frequencies is greater than the number of on and off transitions of telegraphy. Likewise, high-fidelity, wideband FM signals require a greater bandwidth than voice-grade, narrow-band FM systems. Since a large amount of information is conveyed in a conventional fast-scan Television signal, it requires a greater bandwidth than any of the previously mentioned examples. Large objects in a TV picture are transmitted as low frequency information, while small objects and fine detail are transmitted as high frequency information. Fast scanning rates are necessary to portray motion and reduce flickering as the high number of "stills" are integrated into moving pictures. As scanning rates and information content increase, the required RF bandwidth also increases.

Frequency allocations in the RF spectrum are major considerations among major world powers. Each kilohertz of this precious range must be used to its maximum capability and utmost benefit. If fast-scan Television pictures were transmitted in the high-frequency range, worldwide transmissions without the use of satellites would be possible. This would be a drastic misuse of RF spectrum, however, since over 3.5 megahertz would be required to convey the video of each tv channel. A single channel containing video, audio and guard bands would engulf the full range of 7 to 13 MHz. A second channel would annihilate the range of 13 to 19 MHz, etc. Obviously, such thoughts are absurd. Since radio frequencies above approximately 100 MHz traverse line-of-sight distances, this lower-premium spectrum is used for fast-scan TV transmissions. TV signals may be relayed by wideband communications satellites, provided they are situated high enough above earth to attain a line-of-sight position between the specific areas using that satellite.

The advent of slow-scan TV produced a unique innovation to visual communications. These reduced-bandwidth pictures can be transmitted around the world via conventional audio channels such as single sideband or narrow-band FM. Several sacrifices, however, were necessary to accomplish this goal. Video information content was reduced to a minimum acceptable amount, while horizontal and vertical scanning rates were reduced approximately 1000 times. This slowing of rates dictates that all SSTV transmissions consist of a group of "still" pictures rather than conveying motion. A tradeoff of this nature is truly justified when we realize that SSTV permits radio amateurs to actually view distant contacts and share lifestyles on a personal basis. Slow-scan TV has now been accepted on a worldwide scale, and it is providing reliable communications for many remote areas of the world. Thousands of amateurs are expanding their knowledge and complementing international goodwill every day while using this outstanding mode of communication. Such accomplishments were not possible through the medium of audio-only communications. The exclusion of motion in slow-scan TV was a prime inspiration in the attempt to develop a medium-scan TV system. Assuming that an RF bandwidth slightly wider than that used to transmit a conventional SSTV signal could be transmitted in the 29-MHz range of 10-meters, a TV system capable of limited motion and color reproduction could be created. A concept was thus formulated, and plans for a unique long distance motion-TV system began.

BANDWIDTH-REDUCTION TECHNIQUES FOR MSTV

Television in the United States is standardized to what is known as Standard RS-170. This agreed-upon standard, set up by a special committee, has been the reference to which all reduced bandwidth systems have been compared for almost 30 years.

The standard states basically:

- 525 lines per frame
- 60 fields per second (interlaced)
- 30 frames per second

The horizontal resolution of the system is a direct function of the bandwidth of the video or of the modulated RF system. The horizontal resolution is usually measured by the number of alternating black and white vertical lines that can be seen on the viewing monitor. This is sometimes referred to as the number of "pixels" (picture elements) in a digital memory.

In analog television systems such as home television, the resolution is set by the amount of bandwidth in megahertz that is utilized for image reproduction. Our color sets have approximately 3 megahertz of video bandwidth to prevent intermodulation of the 3.58 MHz color burst frequency. The common rule-of-thumb conversion factor for commercial TV is bandwidth is related to resolution by 80 lines per megahertz. This gives 240 lines of horizontal resolution for the average color set.

Any bandwidth-reduction techniques developed will be directed toward the removal of data from this 30-year old accepted standard for television. Let us see how many "degrees of freedom" are left to reduce this TV image from 3 MHz to 1-30 kHz.

The choices are:

- Reduce the number of pixels.
- Reduce the number of lines.
- Reduce the number of fields.
- Intermix some of the above reductions in some acceptable manner.

In the early days of television these same problems were faced for our standards of the present system. In order to reduce the bandwidth to the present 6-MHz total RF-channel bandwidth it was necessary to:

- Transmit 70 Hz fields to eliminate flicker.
- Use interlaced fields to reduce the bandwidth by a factor of 2.
- Us vestigial RF transmission to reduce the bandwidth by an additional factor of 2.

In recent years, there has been one important invention that has changed the above considerations. Movement on the TV screen was automatically arrested when the flicker rate was established at 60 fields per second. With the invention of the digital scan converter it became possible to load a refresh memory at a slower rate than 60 times a second, and still display the image at 60 fields per second.

Tests have shown that under "picturephone" type transmission, 3.75 fields per second (60 divided by 16) are satisfactory to a viewer.

A typical TV-bandwidth calculation can be shown as follows for 3.75 fields per second. The formula is:

$$\text{Fields per second} \times \text{lines per field} \times \text{pixels per line} /2 = \text{BW}$$

$$3.75 \times 128 \times 128/2 = 30 \text{ kHz}$$

Since the base video bandwidth must be half of the RF bandwidth in normal cases, steps must be taken to further reduce the bandwidth to half the 30 kHz. Techniques being demonstrated today are:

- 7.5 Hz field rate with both horizontal and vertical interlace
- 3.75 Hz field rate with either vertical or horizontal interlace
- 1.88 Hz field rate with no interlace

In all cases, the base video bandwidth is approximately 16 kHz. This can be used to FM modulate a high-frequency transmitter. Note that it is assumed no subcarrier will be used. This is different from normal SSTV.

The medium-scan TV RF-transmission method will be direct frequency modulation on the assigned frequency 29.150 MHz. The transmitted bandwidth shall not exceed 36 kHz. It is, of course, necessary to preserve the bandwidth throughout the transmitter and receiver. The direct modulation of any HF transceiver can be done by using a reactance modulator on the VFO. The reception can be on any FM receiver tuning to the correct frequency and having a bandwidth of at least 36 kHz. Old surplus police-radio systems may be used. It is planned, however, that a small, solid-state receiver meeting all the MSTV requirements will be available from Science Workshop here in the United States. Since it will use 10.7 MHz I-F transformers and 455 kHz FM components it will probably be very reasonably priced.

Amateurs desiring to convert their Robot 400 Digital Scan Converters for medium-scan TV will need a second memory board. Boards that will fit into the Robot box are obtainable at cost from W9NTP. The chips can be bought from mail order houses, or, if extreme difficulty is found in obtaining them W9NTP will furnish them at his cost. Please note that in no way are these tests being made to promote commercial sales of anything. The investigators of MSTV feel that hams should be advancing the state of the art, and this is their way of doing it.

PROJECTION OF MSTV GROWTH

Tests conducted at the 1979 Dayton Hamvention definitely confirmed that viewer acceptance *preferred the 2 fields* per second rather than any of the interlace systems demonstrated. The other three 16-kHz systems that were on display were: horizontal interlace at 4 fields per second, vertical interlace at 4 fields per second, and double interlace at 8 fields per second. At 8 fields per second the rate of change was too slow and the eye looked at the break-up rather than the picture as a whole. On the other hand if some way

Table 8-1. Specifications for MSTV System for Completion of FCC STA Two-Year Test Period.

Number of pixels per line	128
Number of lines per field	128
Aspect ratio	1/1
Frame rate	2 F/s
Pixel clock frequency(128 L x 128 P)x F =	32,768 pixels per second.
Line frequency	128 × 2 = 256 Hz
Time of one line	3.91 ms.
Base video bandwidth	32,768/2 = 16,384 Hz.
Horizontal sync. pulse width	.4 ms.
Vertical sync. pulse width	1 line, or 3.91 ms.

could be found to speed up the rate of doing double interlace it would prove very acceptable.

Higher-bandwidth systems were also demonstrated, and everyone liked the 8 fields per second frame-grab system. This, of course, is a 64 kHz system and no easy way can be envisioned to use this rate of transmission. *It was, therefore, concluded that, for the remainder of the Special Temporary Authorization (July 1980) that MSTV operators would employ 2-fields-per-second rates.* A Robot 400 with an extra memory obtainable from W9NTP, and an external small buffer memory, can make it possible for everyone to transmit and to receive limited-motion TV. This is the "First Step Forward." A summary of presently adopted MSTV parameters is shown in Table 8-1, and a block diagram of the system is shown in Fig. 8-1.

After the STA is concluded, a report will be written, and submitted to the FCC along with recommendations for future systems. It is hoped elaborate systems will be granted. This is why each amateur should let it be known when he is interested in getting his technical feet wet.

Let us talk about micro-processor systems. Some of the simpler versions can be built up directly from hardware which doesn't require a particular micro-processor system to become involved.

When we "field-grab" a television picture in 1/60 of a second we hold it in memory until time for the second transmission. This time period will depend on how fast we can get rid of the data in the allocation of bandwidth. For instance, a SSTV image has 128 pixels, 128 lines, and 4 bits per pixel. This gives a total number of 65,536 bits in each stored image. Some time later we store another field and it also has 65,536 bits. We now have two pictures stored in a memory, and they have been placed there at different times. Once we have transmitted the first 65,536 bits is there any need to transmit all 65,536 bits of the second picture? Of course the answer is no. We have assumed the transmission is going to be in digital format, but the principle also holds true if we make an analog transmission.

Books have been written, and Bell Telephone has invested considerable funds searching for the ideal system. Let us take an amateur viewpoint and see how this may be accomplished for a picture-quality image comparable to SSTV. Assume we are using the SSTV format of 128 pixels and 128 lines. Divide the horizontal direction into 16 groups of 8 pixels each. Divide the 128 lines into groups of 8 also. The resulting checkerboard now consists of 256 squares. Each of these squares can be referred to in binary code of 8 bits. Each small square consists of 64 pixels. The total number of bits of information that needs to be transmitted for a given square is 72 bits. This makes the system 89% efficient. In this case, it is assumed that the video data will be transmitted in analog format. If it were possible to transmit it digitally, the system would be even more efficient. The square consists of 256 bits; the total number of bits would be 264. The efficiency is therefore 97%. (Amateur data transmissions were not permitted as this was written).

Before you get too excited, let's consider the problems involved in performing this complicated signal sorting. It is very easy to store a field of digital picture, and compare it to a previously stored field. The comparison can be made with an "exclusive-or" circuit. Each of the little squares can be compared, and a counter used to count the pixel differences or the bit differences inside a given pixel. Assuming that we will compare only pixels, and only the top three bits at that, will give an indication of changes. Once these changes have been counted, the next square can be checked for change. This will take 256 computations. The next step will be to put in order the squares that have the most change. It is very probable that only a few of the squares will have enough changes to make it necessary to transmit them. The time of transmitting a square can be easily calculated: 8 × 8 pixels times the subfield rate equals 16 kHz. The field rate is therefore 512 squares which can be transmitted every second. It is obvious that if the camera is panned, all 256 squares will be changed, and the time required for transmission is 0.5 seconds (2 fields/second). This, of course, is to be expected.

In order to test different algorithms and other techniques, I would like to suggest interim simpler systems before we launch into a full 256 calculation. Consider the screen divided up into three

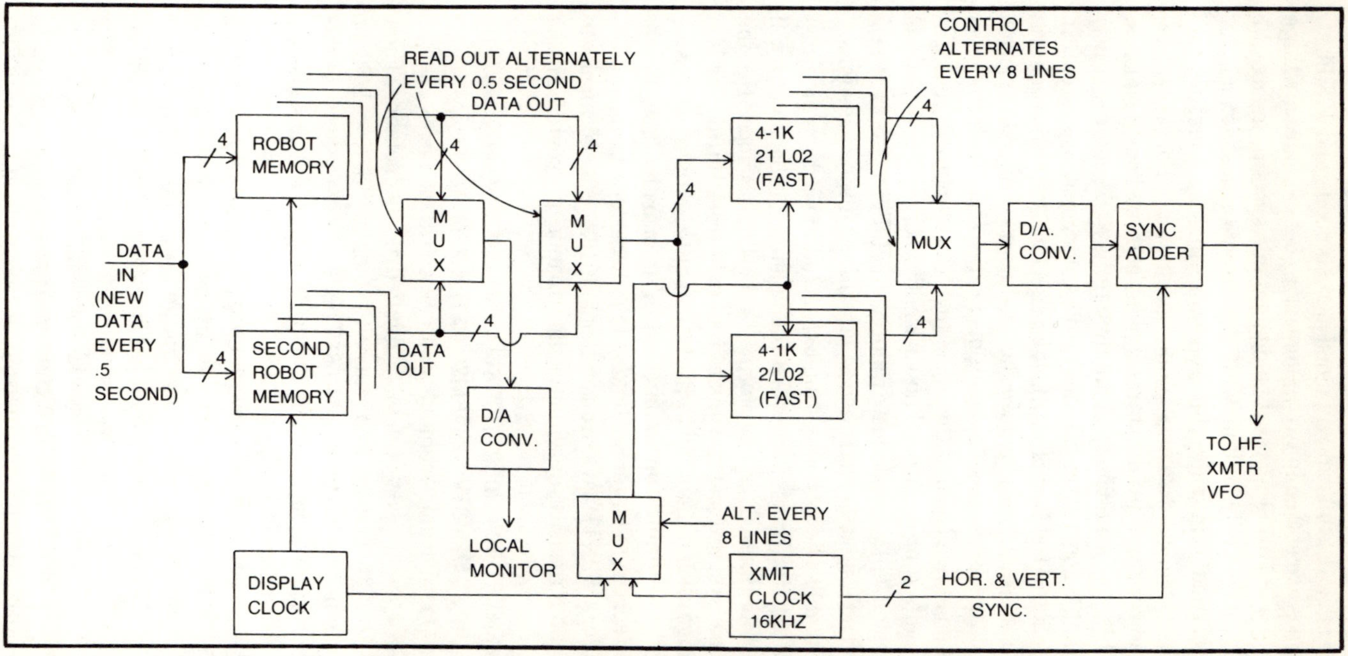

Fig. 8-1. Proposed MSTV system. Pictures are frame-grabbed at 2 fields per second by Robot. This is compatible in both 60 and 50-Hz countries. The Robot memory not being loaded is viewed by the local monitor. During this period of time (0.5 seconds) one of the small static buffer memories is loaded at display rates for 8 lines, while the other buffer is transmitting its 8 lines at 16 kHz.

vertical strips. Each strip will consist of 40 pixels and 128 lines. Two images are stored in two memories as before, but only three calculations and ordering must be done in this system. In fact, the problem can be easily built up with "exclusive-or" and comparator hardware, and eliminate the need for micro-processors and software. Once practice is obtained with a three-area system, divide the vertical lines into three groups and you have 9 squares. The problem now begins to approximate the 256 squares. It may evolve that 9 squares works out to be a very usable system. In review, each of the 9 squares must be checked against its equal in the other stored memory. Once the total number of pixel changes in each of the 9 squares has been made, the ones with the most changes must be placed at the top of the priority list for transmission. At this time it is not known whether or not the speed of transmission will give adequate motion or not. Since MSTV is still in its infancy, future experiments will answer these questions.

All that has been said has been directed toward analog transmission. If digital transmission is contemplated, the bandwidth will increase four times. When this format is chosen, it is hoped that the 256-square system, and the addition of only transmitting the bits that change within the pixel, will allow motion without excessive bandwidth. Additional up-to-date MSTV information may be obtained by sending a large, SASE (2 stamps required) directly to Dr. Don Miller, W9NTP, Waldron, Indiana 46182. Special thanks to Dr. Miller for assisting with the information presented in this chapter.

DIGITAL MEMORY CONSIDERATIONS FOR MEDIUM SCAN MSTV SYSTEMS

A. One dynamic memory plus one static memory:

- Can load a fast-scan picture into dynamic memory. Then load the image into the stitic memory during 1/30 of transmission time. (2 F/S) This won't work because the transmission of data is stopped for 1/30 of transmission time during load.
- On receive it won't work either because of necessity of loading static memory into dynamic memory. Data won't wait during the loading time into dynamic memory.

  Conclusion: *This amount of memory will not permit data-link transmission.*

B. One dynamic memory plus two static memories:

- Can load a fast-scan picture into dynamic memory. Next, load the image into the static memory. Transmit form this static memory while loading the dynamic memory and the other static memory. This will work as far as the transmission is concerned but the display at the sending end will blink during loading of the one dynamic memory.

- On receive, one of the static memories can be loaded. After filling the first static memory the input data is transferred to the other static memory. During this loading time, the first static memory transfers its data into the dynamic display memory. Thepicture will blink during hte write-time unless a gating of vdideo data is made to the monitor during this time.

Conclusion: *System will work, except that on transmit the blinking display leaves something to be desired.*

C. Two dynamic memories plus two static memories:

- Load the dynamic memories alternately on fast scan load. There is no problem with blink since the monitor is alternating between dynamic memories.
- One of the stitic memories is loaded during display time at fast-scan rates. The loaded static memory than transmits. At the end of the transmission, the second fast-scan memory loads the second static memory and the process continues.
- On receive exactly the inverse occurs. One static memory is loaded and then the data is transferred to the dynamic memory. The viewing is from the other dynamic memory so there is no blink.

Conclusion: This system is very satisfactory and lends itself to random loading of the static memories. If, however, the data is being received in a known format it is possible to use less static memory. See next system.

E. Two dynamic memories plus two partial static memories.

- The static memores are 1/16 field, or 4 1-K chips for each static memory.
- The dynamic memories are loaded alternately by fast scan as before.
- The non-viewing dynamic memory is loaded into each of the 1/16 static memories alternately. After 16 partial fields have been alternately transferred to the two partial static memories, the process reverses and the other dynamic memory is used for data transfer.
- On receive, the inverse happens. The two partial static memories are loaded and unloaded alternately. They load

the dynamic memory that is not being viewed so there is no blinking on the screen. This system is very simple, but does not allow for random loading of a complete static memory.

Conclusion: The ultimate system would be to use two dynamic memories plus two static memories. One way that this could be done is to use a microprocessor for the two static memories. The microprocessor may be able to keep up with the data transfer (16 kHz) and use the scan converter for display and frame grab. An alternate way to involve a microprocessor with 32 K of memory would be to have multiplexed addressing and DMA access without going through the CPU.

The discussion about blinking transmit displays can be helped somewhat if the camera is locked to the timing of the memory. This is not the case for the Robot 400, and probably will not be a good option.

What needs to be done is to change a Robot 400 to a two-dynamic-memory system (obtainable from W9NTP and build a static buffer memory from S-100 boards.

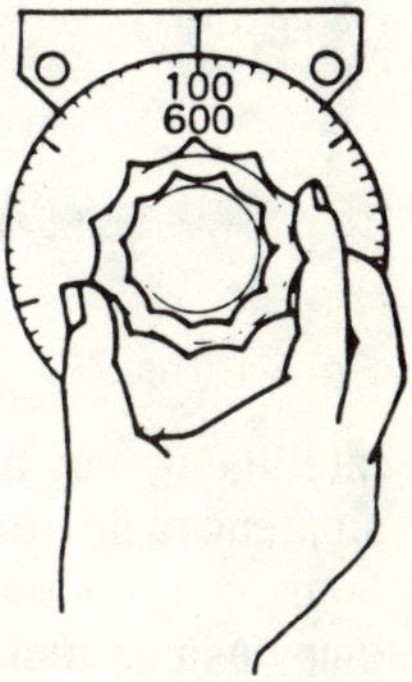

Chapter 9

10 FM Accessories

The vast array of useful accessories available to 10 FM enthusiasts may prove to be a prime asset in this mode's future growth. Since these accessories are available at the present time, it appears 10 FM will not be bound by the development stages experienced on 2-meter FM. The direct compatibility of 10 FM accessories to other amateur modes is another appealing aspect of these units. An amateur purchasing an antenna tuner or noise bridge, for example, will find these times quite beneficial for tuning *all* low-band arrays.

A wide selection of amateur accessories is one of the most logical investments a ham operator can enjoy. These times prove their worth many times over, both at the home station and in the field.

The growing interest in purchasing, rather than home constructing, conventional items such as antenna tuners, CORs, tone encoders, etc., is quite justified. The operation and performance of these times has been established. It's logical to save time by purchasing these units, then apply one's available "construction time" pursuing undeveloped areas. These areas definitely require home construction techniques because commerical equipment simply isn't available. The unique frontier of medium-scan television or direct home-TV satellite terminals are two examples of such "start from scratch" frontiers.

Today's fast-paced living has also brought about another type of amateur operator. This particular individual is usually highly

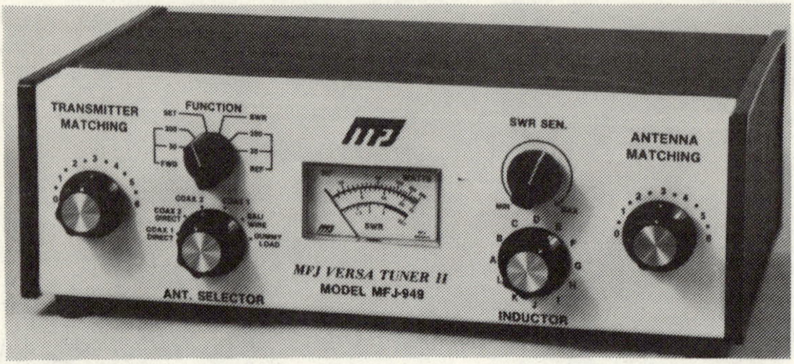

Fig. 9-1. The MFJ Deluxe Versa Tuner.

intelligent, but he has a very limited amount of time to pursue amateur radio interests. If this person manages to allocate time to home construct some basic amateur items, he hasn't time left to enjoy using them. Indeed, this group of amateurs may experience difficulty finding time to enjoy on-the-air operations. Obviously, commercially manufactured accessories are a blessing for these amateurs.

The information presented in this chapter is intended as an overview catalog which 10 FM'ers may use as desired. These manufacturers have favorably established themselves in the amateur world.

ANTENNA TUNERS

MFJ Deluxe Versa Tuner. Allows tuning of antenna over the range of 1.8 to 30 MHz. It has a balun for balanced transmission lines, reads SWR on a dual-range meter, and features a built-in 50-ohm, 200-watt dummy load (Fig. 9-1). Front panel switch selects dummy load, two coaxial lines, direct or through the tuner. The maximum power handling capability is 300 watts. This unit is

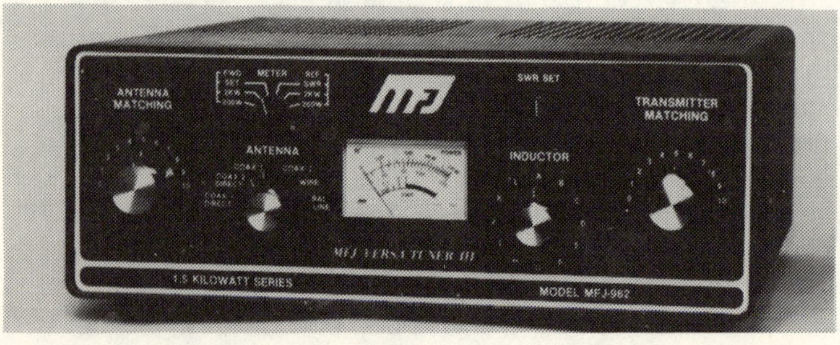

Fig. 9-2. The MFJ Versa Tuner 962.

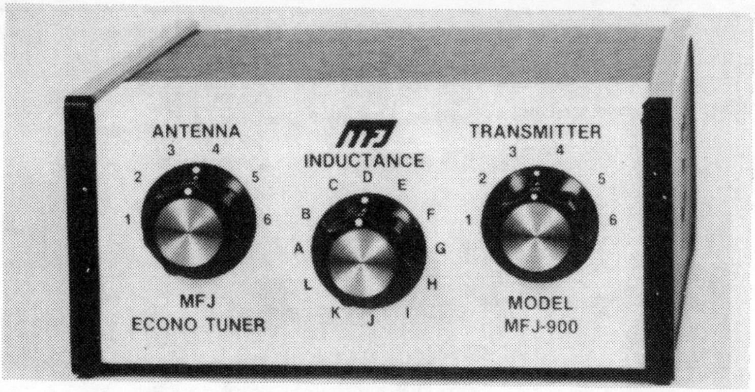

Fig. 9-3. The MFJ Econo Tuner.

available from *MFJ Enterprises, Inc.*, P.O. Box 494, Mississippi State, Mississippi 39762.

MFJ Versa Tuner 962. Handles maximum RF power of 1.5 kW PEP. Matches any one antenna from 1.8 to 30 MHz. Built-in balum; antenna switch selects balanced line or three coax-fed antennas. The mter reads forward ro reflected power, 200 and 2000 watt scales (Fig. 9-2).

MFJ Econo Tuner. Matches coax, random wires, and provides full coverage from 1.8 to 30 MHz (Fig. 9-3). It will handle up to 200 watts. it uses an airwould inductor which gives you less losses than a tapped toroid.

MFJ 16010 Random Wire Tuner. This unit will match random-length wire antennas for coverage of 1.8 to 30 MHz. It will match low impedance antennas by exchanging input for output. It handles up to 200 watts (Fig. 9-4).

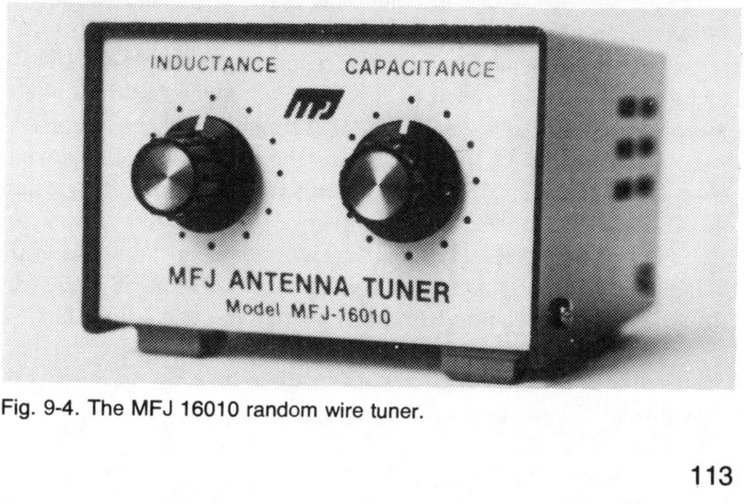

Fig. 9-4. The MFJ 16010 random wire tuner.

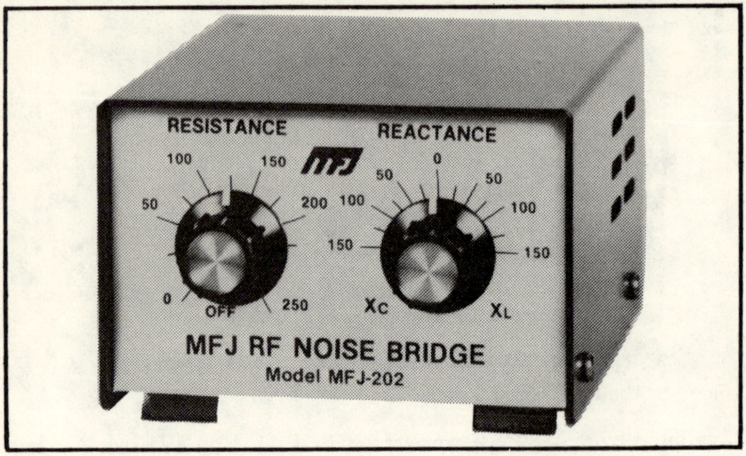

Fig. 9-5. The MFJ RF noise bridge.

MFJ RF Noise Bridge. This model 202 is handy for measuring antenna impedance, resonant frequency, and radiation resistance (Fig. 9-5). It can be used to tune a Transmatch, adjust tuned circuits, measure inductance, etc. Frequency range is 1 to 100 MHz. Resistance range is 0 to 5000 ohms; inductive and capacitive range is 0 to 2200 ohms at 100 MHz.

ANTENNAS

New Tronics Vertical Antenna. The 4BTV antenna covers 10,15, 20 and 40 meters (Fig. 9-6). It's ideal for restricted space and can be made to work on 75 meters with the addition of a resonator. Heavy duty aluminum mounting bracket with low loss-high strength insulators and mounting hardware included. *Specifications*: Length—21′5″; weight—15 lbs.; power capability—full legal limit on SSB and CW; SWR at resonance (typical)—1.15 to 1 or better; bandwidth—1.6 to 1 SWR band edges (40 through 10 meters) ; wind loading—29 lbs. at 70 mph; mounting—vertical support up to 1 ¾′ O.D.; shipping weight—15 lbs. Available from *New Tronics Corp.*, 15800 Commerce Park Dr., Brook Park, Ohio 44142.

New Tronics HOT Mount. This mount is the solution to being able to quickly dismount you antenna and store it in the car trunk. The mount cannot be removed when the trunk lid is closed (Fig. 9-7). It includes 180 degree adjustable swivel for vertical positioning of the antenna. Accepts most flexible antennas with ⅜″ - 24 threaded base up to 55″ in length.

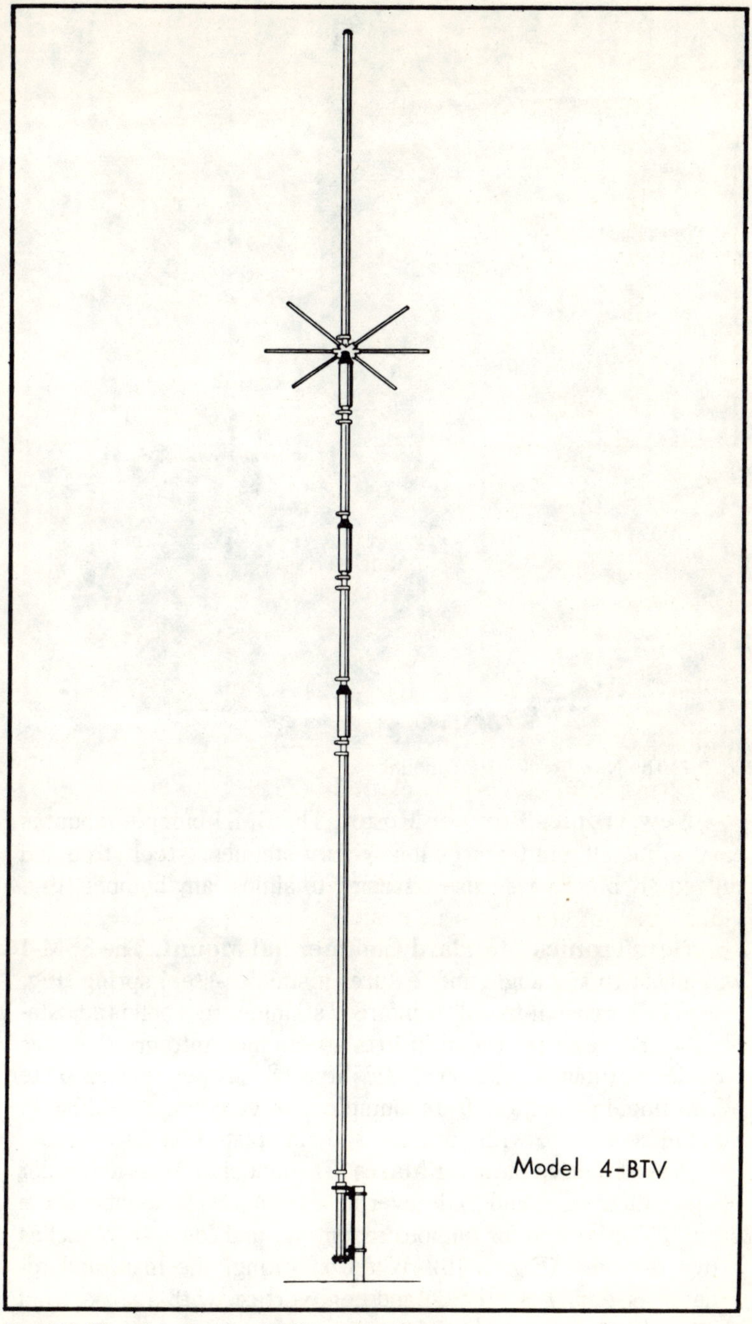

Fig. 9-6. The New Tronics 4BTV vertical antenna.

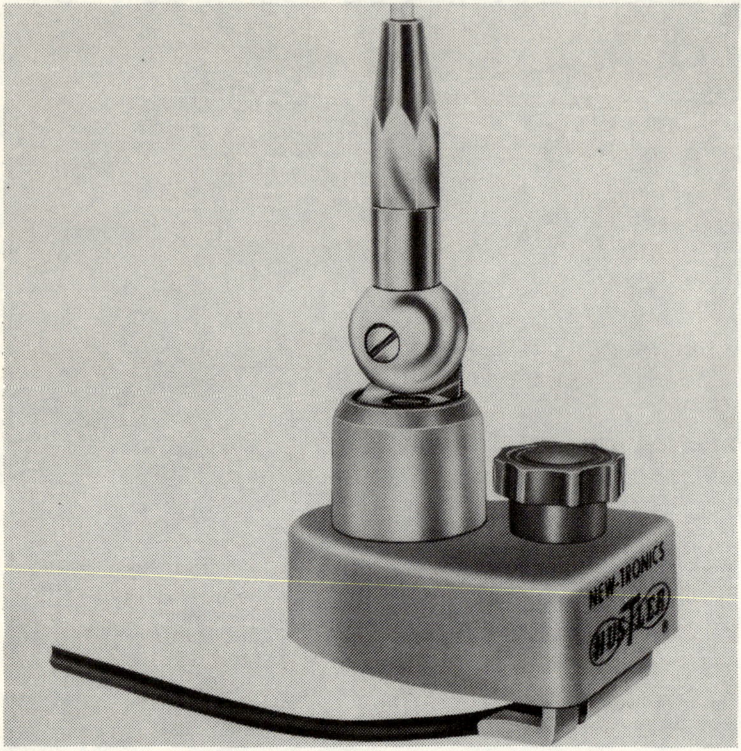

Fig. 9-7. The New Tronics HOT mount.

New Tronics Bumper Mount. The BM-1 bumper mount is easy to install and features heavy-duty stainless-steel strap and rugged "J" bolts for secure fastening to almost any bumper (Fig. 9-8).

New Tronics Standard Commercial Mount. The SSM-1 will adjust to any angle and features a stainles-steel spring (Fig. 9-9). The corrosion-free 2″ reinforced stainless teel ball is adjustable for 180 degrees. Metal inserts are molded into grey cycolac base for optimum mechanical stability. "C" shaped backup plate permitting installation from outside the vehicle is exclusive. Supplied complete with cork pad and zinc plated hardward.

New Tronics Hustler Masts. The model MO-1 is for fender or deck mounting, and folds over at a point 15 inches above the base. The MO-2 is for bumper mounting, and folds at 27 inches above the base (Fig. 9-10). When operating, the mast and resonator assembly is erected and held vertical with a shakeproof sleeve clutch.

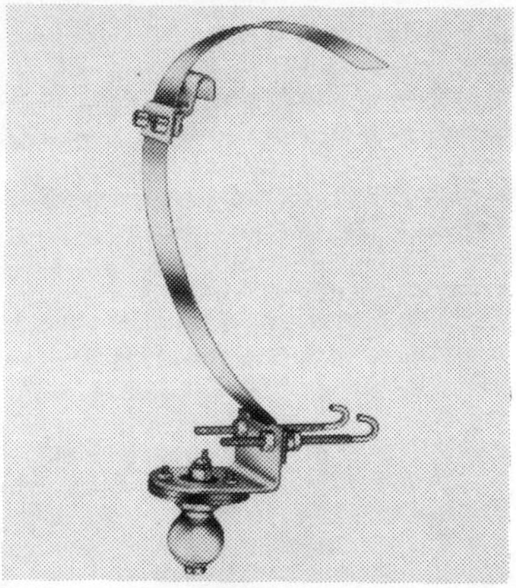

Fig. 9-8. The BM-1 bumper mount from New Tronics.

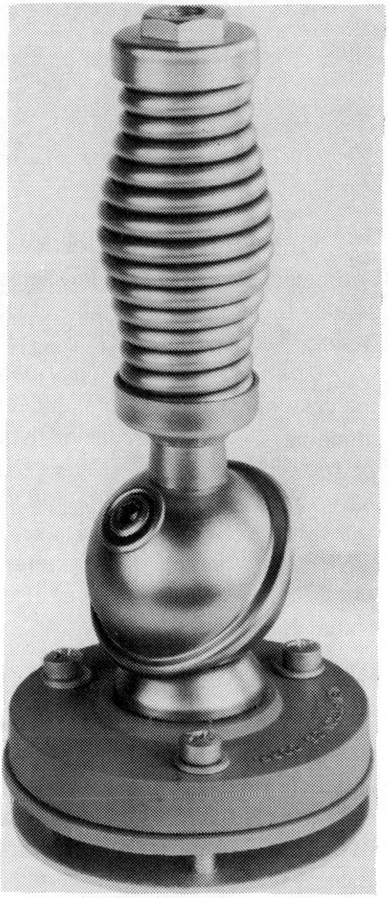

Fig. 9-9. The standard commercial mount SSM-1 from New Tronics.

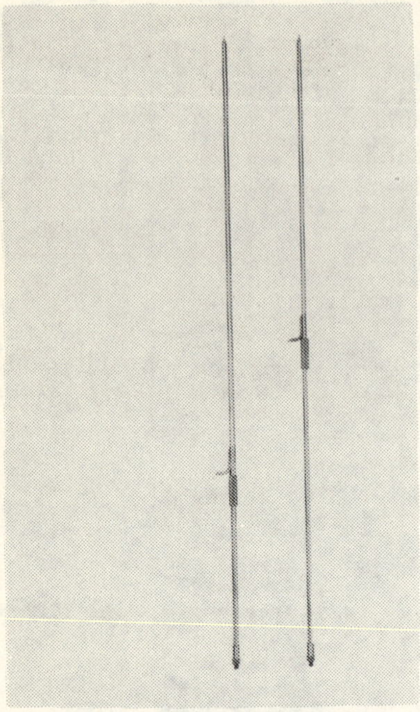

Fig. 9-10. Hustler masts from New Tronics.

New Tronics Hustler Resonators. These resonators are optimized for each band and selaed for weatherproofing. The assembly includes an adjustable stainless steel tip section for precise tuning. The 10-meter RM-10S is shown in Fig. 9-11.

Cushcraft Ringo. The Cushcraft 10FM Ringo bears a striking resemblance to its popular 2-meter counterpart. It is an outstanding ⅝ wave antenna with an omni-directional 3.4 dB-gain pattern. The radiator is 17 feet tall and has a decoupling ring at the base (Fig. 9-12). The tubing sections telescrope for convenient transportation or storage. The antenna can be field-adjusted for low SWR in any part of the 10-meter band. The antenna is avialable from Cushcraft Corporation, P.O. Box 4680, Manchester, New Hampshire 03108.

ENCODERS/DECODERS

Communication Specialists TE-12 Encoder. This encoder (Fig. 9-13) is available for all the commonly used sub-audible and tone-burst frequencies. It can be powered by 6 to 16 volts, DC, and features excellent frequency accuracy and temperature stability. Available from Communications Specialists, P.O. Box 153 Brea, California 92621.

Communications Specialists EM-3 Encoer. This encoder is microminiature in size, only 0.45 × 1.1 × 0.6 inches. It operates on 6-16 volts DC, and has built-in reverse-polarity and voltage protection. It also is immuneto rf interference (Fgi. 9-14).

Communications Specialsists TE-8 Encoder. This unit is compatible with all sub-audible tone systems and is powered by 6-16 volts DC. It uses field-replaceable plug-in K-1 elements for low cost. It is immune to rf and has reverse polarity protection built-in (Fig. 9-15).

Communication Specialists PE-100 Encoder. For a snappy appearance at your home station, Communications Specialists offers this two-twon sequential paging encoder. It is

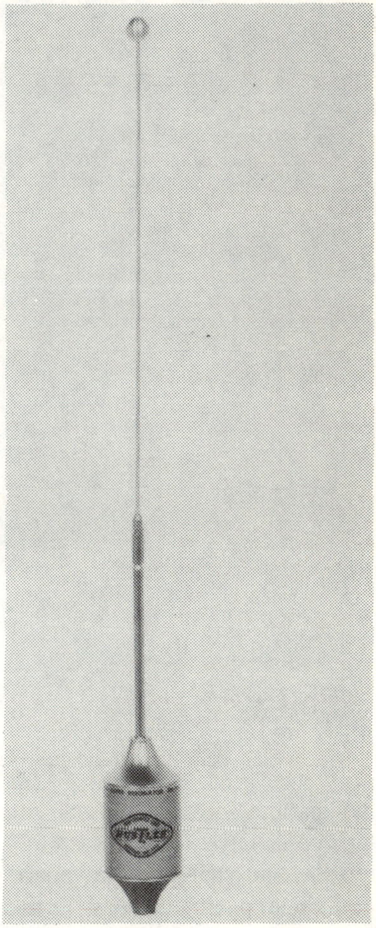

Fig. 9-11. Hustler resonators from New Tronics.

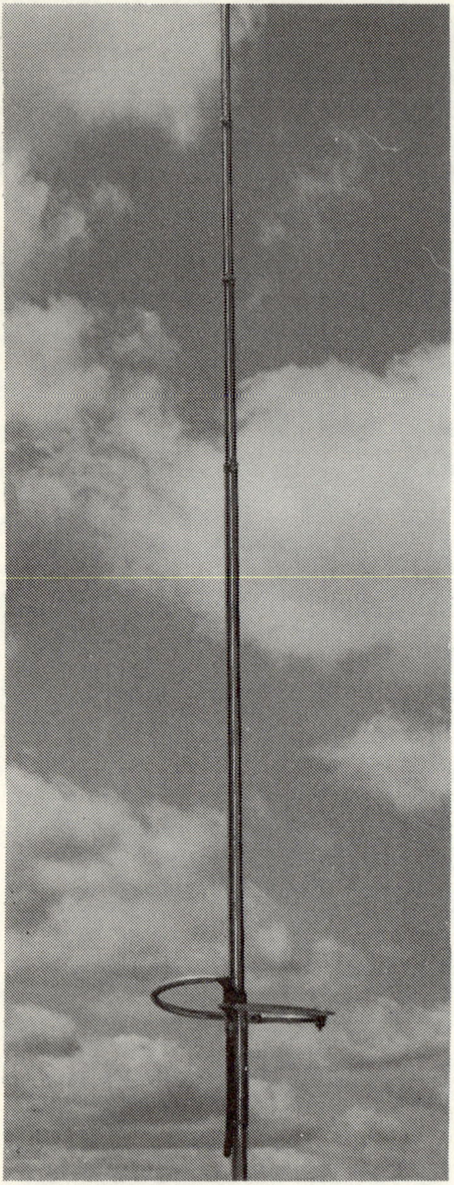

Fig. 9-12. The Cushcraft 10FM Ringo.

compatible with all two-tone sequential systems, and has built-in power supply to work from 120 A AC. It has red LED readout display and a miniature keyboard. It is capable of sending up to 100 individual paging codes (Fig. 9-16).

Fig. 9-13. The Communications Specialists TE-12 encoder.

Communications Specialists SD1 Decoder. The SD-1 two-tone sequential decoder is small enough to fit in all mobile units and most portables (Fig. 9-17). It uses plug-in, field replaceable, K-2 frequency determining elements for low cost. Reverse polarity and over-voltage protection is built-in; also is immune to rf. It may be driven by disciminator, audio stages or speaker, and is powered by 6-16 VDC.

Communications Specialists AP-8 Tone Adapter. This adapter is compatible with all sub-audible tone systems. It adapts

Fig. 9-14. The Communications Specialists ME-3 encoder.

Fig. 9-15. The Communications Specialists TE-8 encoder.

encoder-decoder to eight tone send and receive. Automatic reverting is built-in; its use is optional, to return unit automatically to a common decode frequency during receive (Fig. 9-18).

Fig. 9-16. Communications Specialists PE-100 two-tone sequential paging encoder.

Fig. 9-17. The SD-1 two-tone sequential decoder by Communications Specialists.

Fig. 9-18. The Communications Specialists AP-8 tone adapter.

Fig. 9-19. The DSI 3600A frequency counter.

FREQUENCY COUNTERS

DSI Frequency Counter. The DSI 3600A frequency counter is available in ranges from 50 Hz to 700 MHZ. Its sensitivity is typically 20 mV at 30 MHz, and 10mV up to 250 MHz. (Fig. 9-19). It is available from DSI Instruments, Inc., 7914 Ronson Rd., San Diego, California 92111.

Glossary

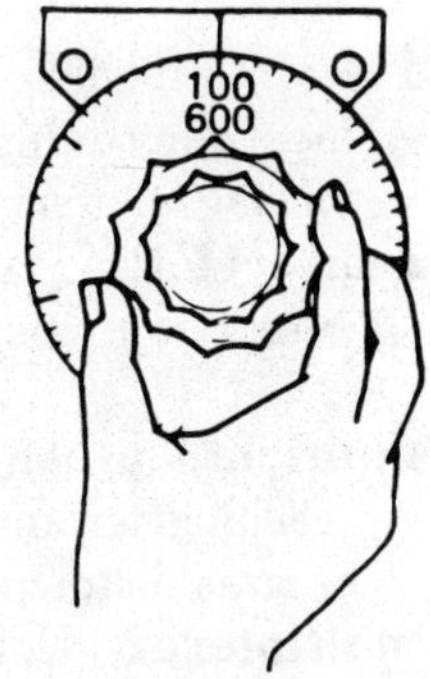

af—audio frequency. This range encompasses 20 Hz to 20 kHz.

barefoot—Term relating to operation of a basic two-way radio, without an external high-power amplifier.

capture effect—The effect of the strongest FM signal capturing a specific frequency while apparently eliminating weaker signals. This unique effect allows one signal to dominate a given frequency at a time.

carrier-operated-relay—A unit which controls a transmitter, activated by squelch voltages from a receiver.

center frequency—The FM signal's measured frequency without modulation, and the FM signal's average frequency with modulation.

decoder—A control circuit which is activated by a specific frequency tone or set of tones. The output of this device is used to control additional devices, such as carrier-operated-relays, telephone links, etc.

direct FM—System where modulation is applied in such a manner as to produce true Frequency Modulation on the output of a modulator stage.

discriminator—Term used for FM detectors. Originally, this term evolved from the Foster-Seeley FM Discriminator.

DX—Term used to describe long-distance communications in amateur radio.

encoder—Unit or circuit which produces a specific frequency tone(s) used for controlling additional units.

I-F—Intermediate Frequency(s) used in communications equipment. Since this frequency is lower than the desired RF range but higher than AF range, it is an intermediate range.

indirect FM—System where phase modulation is produced at the output of a modulator.

multipath propagation—Phenomena that results in a signal reaching an antenna via more than one path. Usually, this creates distortion in FM communications.

multiplexer—Unit or circuit which combines two or more signals into a common output channel.

OSCAR—Acronym for Orbital Satellite Carrying Amateur Radio. These satellites are constructed and placed in orbit under guidance of AMSAT, the Amateur Satellite Corporation. Amateur Satellites operating "Mode A" have their output on the 10 meter band.

remote base—Setup which permits a small, high-frequency unit to operate a larger, high power transceiver which is remotely located. Two Carrier Operated Relays are used for this function. Usually, Remote Base setups are used and maintained by a very small group of Amateurs.

repeater—Setup which receives signals on one frequency and retransmits them on another frequency of the same band. A single transmitter is used in repeater functions. Usually, repeaters are used and maintained by a large group of amateurs.

rf—radio frequencies. These range from approximately 30 kHz to 30 GHz.

sunspot cycle—The approximate 11-year period in which sunspots, or flares, build to a maximum monthly number and decrease to a minimum value. These sunspots directly affect all radio communications.

squelch tail—The time lag between a transmitter's carrier drop and the distant receiver's squelch drop. During this time, a noise burst is heard in the receiver.

Subaudible tone—A control or security tone which is below 300 Hz.

tone access—A security system used with a repeater or remote base to ensure only signals containing the specified tone(s) can utilize that unit.

tone burst—A tone of short duration which is used at the beginning of each transmission to switch on a repeater or remote base. Usually, the frequency and time period of such tones must be maintained within strict tolerances.

VOX—Voice Controlled Transmit (Xmit)—A transmitter-control unit or circuit which is activated by audio on its input circuit. A brief time elapse option is usually included in a VOX to prevent dropout between words.

whistle-access—A repeater-control scheme which requires a brief whistle on its input frequency to activate the system. Also, another name for tone burst access.

Appendix

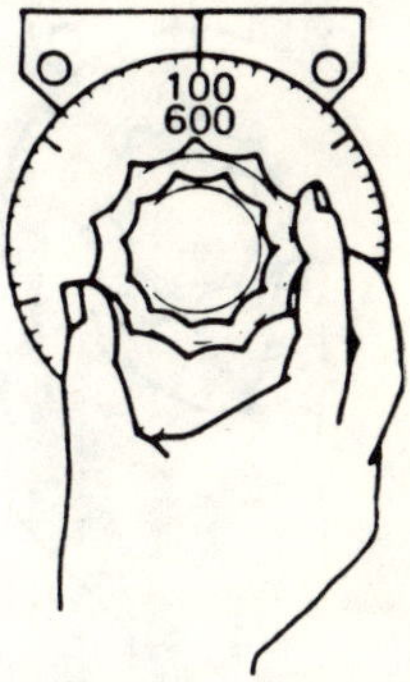

Appendix A
Phonetic Alphabet

A ALFA
B BRAVO
C CHARLIE
D DELTA
E ECHO
F FOXTROT
G GOLF
H HOTEL
I INDIA
J JULIETTE
K KILO
L LIMA
M MIKE
N NOVEMBER
O OSCAR
P PAPA
Q QUEBEC
R ROMEO
S SIERRA
T TANGO
U UNIFORM
V VICTOR
W WHISKEY
X X-RAY
Y YANKEE
Z ZULU

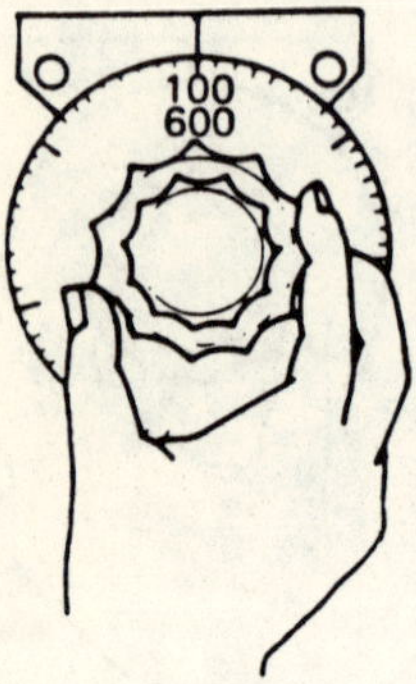

Appendix B
International "Q" Signals

QRA What is the name of your station?
QRB How far are you from my station?
QRD Where are you going and where are you from?
QRG Will you tell me my exact frequency?
QRH Does my frequency vary?
QRI How is the tone of my transmission?
QRK What is the readability of my signals?
QRL Are you busy?
QRM Are you being interferred with?
QRN Are you troubled by static?
QRO Shall I increase power?
QRP Shall I decrease power?
QRQ Shall I send faster?
QRR Are you ready for automatic operation?
QRS Shall I send more slowly?
QRT Shall I stop sending?
QRU Have you anything for me?
QRV Are you ready?
QRW Shall I inform ______ that you are calling him on__kHz.
QRX When will you call me again?
QRY What is my turn?
QRZ Who is calling me?
QSA What is the strength of my signals?
QSB Are my signals fading?
QSD Is my keying defective?
QSG Shall I send ______ telegrams at a time?
QSJ What is the charge to be collected per word to ______ including your internal telegraph charge?
QSK Can you hear me between your signals?
QSL Can you acknowledge receipt?
QSM Shall I repeat the last telegram which I sent you, or some previous telegram?
QSN Did you hear me on ______ kHz?
QSO Can you communicate with ______ direct or by relay?
QSP Will you relay to ______ free of charge?
QSQ Have you a doctor on board [or is ... (name of person) on board]?
QSU Shall I send or replay on this frequency [or on ______ kHz] (with emission of class______)?
QSV Shall I send a series of V's on this frequency?
QSW Will you send on this frequency?
QSX Will you listen to ______ on ______ kHz?
QSY Shall I change to transmission on another frequency?
QSZ Shall I send each word or group more than once?
QTA Shall I cancel telegram number ______ as if it had not been sent?
QTB Do you agree with my counting of words?
QTC How many telegrams have you to send?
QTE What is my true bearing from you?
QTG Will you send two dashes of ten seconds each followed by your call sign [on ______ kHz]?
QTH What is your location?
QTI What is your true track?
QTJ What is your speed?
QTL What is your true heading?
QTN At what time did you leave ______ (place)?
QTO Have you left dock (or port)?
or
Are you airborne?
QTP Are you going to enter dock (or port)?
or
Are you going to land?
QTQ Can you communicate with my station by means of the International Code of Signals?
QTR What is the correct time?
QTS Will you send your call sign for ______ minute(s) on ______ kHz so that your frequency may be measured?
QTU What are the hours during which your station is open?
QTV Shall I stand guard for you on the frequency of ______ kHz?
QTX Will you keep your station open for further communication with me until further notice?
QUA Have you news of ______ (call sign)?
QUB Can you give me, in the following order, information concerning: visibility, height of clouds, direction and velocity (place of observation)?
QUC What is the number of the last message you received from me?
QUD Have you received the urgency signal, sent by ______ (call sign of mobile station)?
QUF Have you received the distress signal sent by ____ (call sign of mobile station)?
QUG Will you be forced to land?
QUH Will you give me the present barometric pressure at sea level?

Appendix C

International Prefixes

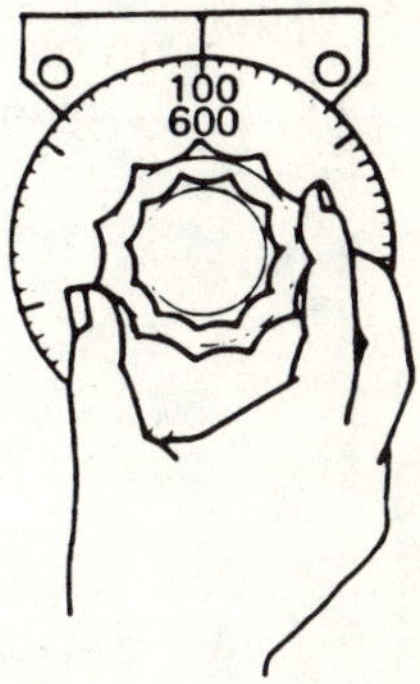

PREFIX	COUNTRY
AP	PAKISTAN
A2C	BOTSWANA
A35	TONGA
A4X	SULTANATE OF OMAN
A51	BHUTAN
A6X	UNITED ARAB EMIRATES
A7X	QATAR
A9X	BAHRAIN
BV	TAIWAN
BY	PEOPLES REPUBLIC OF CHINA
CE	CHILE
CE9AA-AM	CHILEAN ANTARCTICA TIERRA DE O'HIGGINS, PALMER PENINSULA, GRAHAM LAND
CE9AN-AZ	SOUTH SHETLAND IS.
CE0A	EASTER IS.
CE0X	SAN FELIX IS.
CE0Z	JUAN FERNANDEZ IS.
CM	CUBA
CN	MOROCCO
CO	CUBA
CP	BOLIVIA
CR3	GUINEA BISSAU
CR9	MACAO
CT1	PORTUGAL
CT2	AZORES IS.
CT3	MADEIRA IS.
CX	URUGUAY
C21	REPUBLIC OF NAURU
C31	ANDORRA
C5A	GAMBIA
C6A	BAHAMA IS.
C9M	MOZAMBIQUE
DA-DL	FED REP OF WEST GERMANY
DM	EAST GERMAN DEMOCRATIC REP.
DU	PHILIPPINES
D2A	ANGOLA
D4	REP OF CAPE VERDE
D6	STATE OF COMORO
EA	SPAIN
EA6	BALEARIC IS.
EA8	CANARY IS.
EA9	SPANISH SAHARA, CEUTA Y MELILLA
EI	IRELAND
EL,5L	LIBERIA
EP	IRAN
ET3	ETHIOPIA
F	FRANCE
FB8W	CROZET IS.
FB8X	KERGUELEN IS.
FB8Y	ANTARCTICA
FB8Z	AMSTERDAM & ST. PAUL IS.
FC	CORSICA
FG7	GUADELOUPE
FH8	MAYOTTE
FK8	NEW CALEDONIA
FM7	MARTINIQUE
FO8	FRENCH POLYNESIA, CLIPPERTON IS.
FP8	ST. PIERRE & MIQUELON IS.
FR7	GLORIOSO, JUAN DE NOVA, REUNION, TROMELIN IS.
FS7	ST. MARTIN IS.
FW8	WALLIS & FUTUNA IS.
FY7	FRENCH GUIANA
G	ENGLAND
GD	ISLE OF MAN
GI	NORTHERN IRELAND
GJ	JERSEY
GM	SCOTLAND
GU	BAILIWICK OF GUERNSEY
GW	WALES
HA,HG	HUNGARY
HB	SWITZERLAND
HB0	LIECHTENSTEIN
HC	ECUADOR
HC8	GALAPAGOS IS.
HG	HUNGARY
HH	HAITI
HI	DOMINICAN REPUBLIC
HK	COLOMBIA

HK0..BAJO NUEVO,MALPELO,SAN ANDRES & PROVIDENCIA IS.
HK0........SERRANA BANK & RONCADOR CAY
HM,HL9........KOREA
HP........PANAMA
HR........HONDURAS
HR0........SWAN IS.
HS........THAILAND
HV........VATICAN CITY
HZ,7Z........SAUDI ARABIA
H4........SOLOMON IS.

I, IW........ITALY
IA........TUSCAN ARCHIPELAGO
IC........CAPRI & ISCHIA IS.
IG9........LAMPEDUSA IS.
IH........PANTELLERIA IS.
IM........MADDALENA IS.
IS........SARDINIA
IT........SICILY

JA,JE-JJ,JR........JAPAN
JD1.......OGASAWARA,MINAMI-TORI-SHIMA IS.
JR6........OKINAWA (RYUKYU IS.)
JT1........MONGOLIA
JW........SVALBARD IS.
JX........JAN MAYEN IS.
JY........JORDAN
J2........REP OF DJIBOUTI
J3........GRENADA & DEPENDENCIES

K........UNITED STATES OF AMERICA
KA........U.S.PERSONNEL IN JAPAN
KB6........BAKER,CANTON,ENDERBURY, HOWLAND & PHOENIX IS.
KC4........NAVASSA IS.
KC4AA,KC4US........ANTARCTICA
KC6........CAROLINE IS.
KG4........GUANTANAMO BAY
KG6........MARIANA IS.
KG6........GUAM
KG6R........ROTA
KG6S........SAIPAN
KH6........HAWAII & KURE IS.
KJ6........JOHNSTON IS.
KL7........ALASKA
KM6........MIDWAY IS.
KP4........PUERTO RICO
KP6........JARVIS & PALMYRA IS.
KS6........AMERICAN SAMOA
KV4........VIRGIN IS.
KW6........WAKE IS.
KX6........MARSHALL IS.
KZ5........CANAL ZONE

LA-LJ........NORWAY
LU........ARGENTINA
LU-Z........ANTARCTICA
LX........LUXEMBOURG
LZ........BULGARIA

M1........SAN MARINO

N........UNITED STATES OF AMERICA

OA........PERU
OD5........LEBANON
OE........AUSTRIA
OH........FINLAND
OH0........ALAND IS.
OJ0, OH0M........MARKET REEF
OK,OL........CZECHOSLOVAKIA
ON........BELGIUM
OR........ANTARCTICA
OX........GREENLAND
OY........FAEROES IS.
OZ........DENMARK

PA-PI........NETHERLANDS
PJ........NETHERLANDS ANTILLES
PJ2,9........CURACAO
PJ3,9........ARUBA
PJ4,9........BONAIRE
PJ5,8........ST. EUSTATIUS
PJ6,8........SABA IS.
PJ7,8........SINT MAARTEN
PP-PY........BRAZIL
PY0........FERNANDO DE NORONHA IS.
PY0........ST. PETER & ST. PAUL'S ROCKS
PY0........TRINIDADE & MARTIM VAZ IS.
PZ........SURINAM
P29........PAPUA NEW GUINEA

SJ-SM........SWEDEN
SP........POLAND
ST........SUDAN
SU........EGYPT
SV........CRETE,GREECE
SV5........DODECANESE IS.
S2,S3........BANGLADESH
S7........SEYCHELLES IS.
S8........TRANSKEI
S9........SAO TOME & PRINCIPE IS.

TA,TC........TURKEY
TF........ICELAND
TG........GUATEMALA
TI........COSTA RICA
TI9........COCOS IS.
TJ........CAMEROON
TL8........CENTRAL AFRICAN REPUBLIC
TN8........REPUBLIC OF CONGO
TR8........GABON REPUBLIC
TT8........REPUBLIC OF CHAD
TU2........IVORY COAST
TY........PEOPLLES REPUBLIC OF BENIN
TZ........MALI REPUBLIC

UA1,2,3,4,6........EUROPEAN RUSSIAN SOVIET FEDERATED SOCIALIST REPUBLIC
UA9,0........ASIATIC RUSSIAN S.F.S.R.
UA1........FRANZ JOSEF LAND
UA2........KALININGRADSK
UB5........UKRAINIAN S.S.R.
UC2........WHITE RUSSIAN S.S.R.
UD6........AZERBAIDZHAN S.S.R.
UF6........GEORGIAN S.S.R.
UG6........ARMENIAN S.S.R.
UH8........TURKMEN S.S.R.
UI8........UZBEK S.S.R.
UJ8........TADZHIK S.S.R.
UL7........KAZAKH S.S.R.
UM8........KIRGHIZ S.S.R.
UN1........KARELO-FINNISH S.S.R.
UO5........MOLDAVIAN S.S.R.
UP2........LITHUANIAN S.S.R.
UQ2........LATVIAN S.S.R.
UR2........ESTONIAN S.S.R.

VE........CANADA
VK........AUSTRALIA
VK2........LORD HOWE IS.
VK9N........NORFOLK IS.
VK9X........CHRISTMAS IS.
VK9V........COCOS IS.
VK9Z........WILLIS IS.
VK0........ANTARCTICA
VO1........NEWFOUNDLAND
VO2........LABRADOR
VP1........BELIZE
VP2........LEEWARD & WINDWARD IS.
VP2A........ANTIGUA,BARBUDA
VP2O........DOMINICA
VP2E........ANGUILLA
VP2K........ST KITTS,NEVIS
VP2L........ST LUCIA
VP2M........MONSTERRAT
VP2S........ST VINCENT & DEPENDENCIES
VP2V........BRITISH VIRGIN IS.
VP5........TURKS & CAICOS IS.

VP8 ... FALKLAND,S.GEORGIA,S.ORKNEY, S.SANDWICH,S.SHETLAND IS.,GRAHAM LAND
VP9 ... BERMUDA IS.
VQ9 ... CHAGOS
VR1 ... BRITISH PHOENIX,GILBERT & OCEAN IS.
VR3 ... NORTHERN LINE IS.
VR6 ... PITCAIRN IS.
VR7 ... CENTRAL & SOUTHERN LINE IS.
VR8 ... TUVALU IS.
VS5 ... BRUNEI
VS6 ... HONG KONG
VU2 ... INDIA
  VU7 ... ANDAMAN & NICOBAR IS.
  VU7 ... LACCADIVE IS.

W ... UNITED STATES OF AMERICA

XE,XF ... MEXICO
  XF4 ... REVILLA GIGEDO IS.
XT2 ... VOLTAIC REPUBLIC
XU ... CAMBODIA/KHMER REPUBLIC
XV5 ... VIETNAM
XW8 ... LADS
XZ ... BURMA

YA ... AFGHANISTAN
YB-YD ... INDONESIA,TIMOR IS.
YI ... IRAQ
JQ ... NEW HEBRIDES
YK ... SYRIA
YN ... NICARAGUA
YO ... ROMANIA
YS ... EL SALVADOR
YU ... YUGOSLAVIA
YV ... VENEZUELA
  YVO ... AVES IS.

ZA ... ALBANIA
ZB2 ... GIBRALTAR
ZD7 ... SAINT HELENA IS.
ZD8 ... ASCENSION IS.
ZD9 ... TRISTAN DA CUNHA & GOUGH IS.
ZE ... RHODESIA
ZF1 ... CAYMAN IS.
ZK1 ... COOK & MANIHIKI IS.
ZK2 ... NIUE IS.
ZL ... NEW ZEALAND & AUCKLAND,CAMPBELL,CHATHAM,KERMADEC IS.
  ZL5 ... ANTARCTICA
ZM7 ... TOKELAU IS.
ZP ... PARAGUAY
ZR,ZS1,2,4,5,6 ... REPUBLIC OF SOUTH AFRICA
  ZS1ANT ... ANTARCTICA
  ZS2 ... PRINCE EDWARD & MARION IS.
ZR,ZS3 ... SOUTHWEST AFRICA (NAMIBIA)

3A ... MONACO
3B6 ... AGALEGA IS.
3B7 ... ST. BRANDON IS.
3B8 ... MAURITIUS IS.
3B9 ... RODRIGUEZ IS.

3C ... EQUATORIAL GUINEA
3D2 ... FIJI IS.
3D6 ... SWAZILAND
3V8 ... TUNISIA
3X ... REPUBLIC OF GUINEA
3Y ... BOUVET IS.

4K1 ... ANTARCTICA
4S7 ... SRI LANKA
4U ... UNITED NATIONS,GENEVA
4W ... YEMEN
4X4,4Z4 ... ISRAEL

5A ... LIBYAN ARAB REPUBLIC
5B4,ZC4 ... CYPRUS
5HI ... ZANZIBAR,TANZANIA
5H3 ... TANZANIA
5L ... LIBERIA
5N2 ... NIGERIA
5R8 ... MALAGASY REPUBLIC
5T5 ... MAURITANIA
5U7 ... NIGER
5V ... TOGO
5W1 ... WESTERN SAMOA
5X5 ... UGANDA
5Z4 ... KENYA

6O ... SOMALI REPUBLIC
6W8 ... SENEGAL REPUBLIC
6Y5 ... JAMAICA

7J ... OKINO-TORI-SHIMA
7O ... SOUTH YEMEN & KAMARAN IS
7P8 ... LESOTHO
7Q7 ... MALAWI
7X ... ALGERIA
7Z ... SAUDI ARABIA

8J ... ANTARCTICA
8P6 ... BARBADOS IS.
8Q6 ... MALDIVE IS.
8R ... GUYANA
8Z4 ... SAUDI ARABIA/IRAQ NEUTRAL ZONE

9G1 ... GHANA
9H1,5 ... MALTA
  9H4 ... GOZO (MALTA)
9I,9J ... ZAMBIA
9K2 ... KUWAIT
9L1 ... SIERRA LEONE
9M2 ... WEST MALAYSIA
9M6 ... SABAH
9M8 ... SARAWAK
9N1 ... NEPAL
9Q5 ... REPUBLIC OF ZAIRE
9U5 ... BURUNDI
9V1 ... REPUBLIC OF SINGAPORE
9X ... RWANDA
9Y4 ... TRINIDAD & TOBAGO IS.

COUNTRY	*PREFIX*
AFGHANISTAN	YA
AGALEGA IS	3B6
ALAND IS	OHO
ALASKA	KL7
ALBANIA	ZA
ALGERIA	7X
AMSTERDAM & ST. PAUL IS	FB8Z
ANDAMAN & NICOBAR IS	VU7
ANDORRA	C31
ANGOLA	D2A
ANTARCTICA	CE9AA-AM,FBBY,KC4,LU-Z, UA1,VKO,VP8,ZL5,ZS1ANT,3Y,4K1,8J
ARGENTINA	LU
ARUBA	PJ3,9
ASCENSION IS	ZD8
AUCKLAND & CAMPBELL IS	ZL
AUSTRALIA	VK
AUSTRIA	OE
AVES IS	YVO
AZDRES IS	CT2
BAHAMA IS	C6A
BAHRAIN IS	A9X
BAJO NUEVO IS	HKO
BAKER IS	KB6
BALEARIC IS	EA6
BANGLADESH	S2,S3
BARBADOS IS	8P6
BELGIUM	ON
BELIZE	VPI
BENIN, PEOPLE'S REP OF	TY
BERMUDA IS	VP9
BHUTAN	A51
BOLIVIA	CP
BONAIRE	PJ4,9
BOTSWANA	A2C

BOUVET IS ... 3Y
BRAZIL ... PP-PY
BRITISH PHOENIX IS ... VRI
BRUNEI ... VS5
BULGARIA ... LZ
BURMA ... XZ
BURUNDI ... 9U5

CAMBODIA/KHMER REP ... XU
CAMEROON ... TJ
CANADA ... VE
CANAL ZONE ... KZ5
CANARY IS ... EA8
CANTON IS ... KB6
CAPE VERDE, REP OF ... D4
CAPRI IS ... IC
CAROLINE IS ... KC6
CAYMAN IS ... ZF1
CENTRAL AFRICAN REPUBLIC ... TL8
CEUTA Y MELILLA, SPANISH ... EA9
CHAD REPUBLIC ... TT8
CHAGOS IS ... VQ9
CHATHAM IS ... ZL
CHILE ... CE
CHINA, PEOPLES REP OF ... BY
CHRISTMAS IS ... VK9X
CLIPPERTON IS ... FO8
COCOS IS ... TI9
COCOS (KEELING) IS ... VK9Y
COLUMBIA ... HK
COMORO,STATE OF ... D6
CONGO, REPUBLIC OF ... TN8
COOK IS ... ZK1
CORSICA ... FC
COSTA RICA ... TI
CRETE, GREECE ... SV
CROZET IS ... FB8W
CUBA ... CM,CO
CURACAO ... PJ2,9
CYPRUS ... 5B4,ZC4
CZECHOSLOVAKIA ... OK,OL

DENMARK ... OZ
DJIBOUTI,REP OF ... J2
DODECANESE IS ... SV5
DOMINICAN REPUBLIC ... HI

EASTER IS ... CE0A
ECUADOR ... HC
EGYPT ... SU
EL SALVADOR ... YS
ENDERBURY IS ... KB6
ENGLAND ... G
ETHIOPIA ... ET3

FAEROES IS ... DY
FALKLAND IS ... VP8
FERNANDO DE NORONHA ... PY0
FIJI IS ... 3D2
FINLAND ... OH
FRANCE ... F
FRENCH POLYNESIA ... FO8

GABON REPUBLIC ... TR8
GALAPAGOS IS ... HC8
GAMBIA ... C5A
GERMANY, FED. REP. (WEST) ... DA-DL
GERMAN DEM. REP. EAST ... DM
GHANA ... 9G1
GIBRALTAR ... ZB2
GILBERT & OCEAN IS ... VRI
GLORIOSO IS ... FR7
GOUGH IS ... ZD9
GOZO (MALTA) ... 9H4
GRAHAM LAND ... VP8,LU-Z
GREECE ... SV
GREENLAND ... OX
GRENADA ... J3
GUADELOUPE ... FG7
GUAM IS ... KG6

GUANTANAMO BAY ... KG4
GUATEMALA ... TG
GUERNSEY,BAILIWICK OF ... GU
GUIANA, FRENCH ... FY7
GUINEA BISSAU ... CR3
GUINEA, EQUATORIAL ... 3C
GUINEA, REPUBLIC OF ... 3X
GUYANA ... 8R

HAITI ... HH
HAWAII ... KH6
HEARD IS ... VK0
HONDURAS ... HR
HONG KONG ... VS6
HOWLAND IS ... KB6
HUNGARY ... HA, HG

ICELAND ... TF
INDIA ... VU2
INDONESIA ... YB-YD
IRAN ... EP
IRAQ ... Y1
IRELAND ... E1
ISCHIA ... IC
ISLE OF MAN ... GD
ISRAEL ... 4X4,4Z4
ITALY ... I,IW
IVORY COAST ... TU2

JAMAICA ... 6Y5
JAN MAYEN IS ... JX
JAPAN ... JA-JR
JAPAN, U.S.PERSONNEL IN ... KA
JARVIS IS ... KP6
JERSEY ... GJ
JOHNSTON IS ... KJ6
JORDON ... JY
JUAN DE NOVA IS ... FR7
JUAN FERNANDEZ IS ... CE02

KAMARAN IS ... 7D
KENYA ... 5Z4
KERGUELEN IS ... FB8X
DERMADEC IS ... ZL
KOREA ... HM,HL9
KURE IS ... KH6
KUWAIT ... 9K2

LABRADOR ... VD2
LACCADIVE IS ... VU7
LAMPEDUSA IS ... IG9
LAOS ... XW8
LEBANON ... OD5
LEEWARD IS.
- ANGUILLA ... VPZE
- ANTIGUA, BARBUDA ... VP2A
- BRITISH VIRGIN IS ... VP2V
- MONTSERRAT ... VP2M
- ST. KITTS, NEVIS ... VP2K

LESOTHO ... 7P8
LIBERIA ... EL,5L
LIBYAN ARAB REPUBLIC ... 5A
LIECHTENSTEIN ... HB0
LINE IS, NORTHERN ... VR3
LINE IS,CENTRAL & SOUTHERN ... VR7
LORD HOWE IS ... VK2
LUXEMBOURG ... LX

MACAO ... CR9
MACQUARIE IS ... VK0
MADDALENA IS ... IM
MADEIRA IS ... CT3
MALAGASY REPUBLIC ... 5R8
MALAWI ... 7Q7
MALAYSIA, EAST ... 9M6,8
MALAYSIA, WEST ... 9M2
MALDIVE IS ... 8Q6
MALI REPUBLIC ... TZ
MALPELO IS ... HK0
MALTA ... 9HI,5
MANIHIKI IS ... 2K1

MARIANA IS KG6
MARKET REEF 0J0,0H0M
MARSHALL IS KX6
MARTINIQUE FM7
MAURITANIA 5T5
MAURITIUS IS 3B8
MAYOTTE FH8
MEXICO ..... XE.AF
MIDWAY IS KM6
MINAMI-TORI-SHIMA IS JD1
MONACO 3A
MONGOLIA JT1
MOROCCO 0N
MOZAMBIQUE 09M

NAURU, REPUBLIC OF Q21
NAVASSA IS KC4
NEPAL 9N1
NETHERLANDS ..... PA-PI
NETHERLANDS ANTILLES ..... PJ
NEW CALEDONIA FK8
NEWFOUNDLAND V01
NEW HEBRIDES ..... YJ
NEW ZEALAND ..... ZL
NICARAGUA ..... YN
NIGER 5U7
NIGERIA 5N2
NIUE IS ZK2
NORFOLK IS VK9N
NORTHERN IRELAND ..... GI
NORWAY ..... LA-LJ

OGASAWARA IS ..... JDI
OKINAWA (RYUKYU IS.) JR6
OKINO-TORI-SHIMA 7J1
OMAN, SULTANATE OF A4K

PAKISTAN ..... AP
PALMYRA IS KP6
PANAMA ..... HP
PANTELLERIA IS ..... IH
PAPUA NEW GUINEA P29
PARAGUAY ..... ZP
PERU 0A
PHILLIPINES ..... DU
PHEONIX IS K36
PITCAIRN IS VR6
POLAND ..... SP
PORTUGAL CT1
PRINCE EDWARD & MARION IS. Z52
PRINCIPE & SAO TOME IS S9
PUERTO RICO KP4

QATAR A7K

REUNION IS FR7
REVILLA GIGEDO IS XF4
RHODESIA ..... ZE
RODRIGUEZ IS 3B9
ROMANIA Y0
ROTA KG6R
RWANDA 9X

SABAH 9M6
SABA IS PJ6,8
ST. BRANDON IS 3B7
ST. EUSTATIUS PJ5,8
SAINT HELENA IS ZD7
SAINT MARTIN IS FS7
ST. PETER & ST. PAULS ROCKS PY0
ST. PIERRE & MIQUELON IS FP8
SAIPAN KG6S
SAMOA, AMERICAN KS6
SAMOA, WESTERN 5WI
SAN ANDRES & PROVIDENCIA HK0
SAN FELIX IS CE0X
SAN MARINO M1
SARAWAK 9M8
SARDINIA ..... IS
SAUDI ARABIA ..... HZ, 7Z
SAUDI ARABIA/IRAQ
NEUTRAL ZONE 8Z4
SCOTLAND ..... GM
SENEGAL REPUBLIC 6W8
SERRANA BK & RONCADOR CAY 8K0
SEYCHELLES IS S79
SICILY ..... IT
SIERRA LEONE 9LI
SINGAPORE, REP OF 9V1
SINT MAARTEN PJ7,8
SOLOMON IS H4
SOMALI REPUBLIC 60
SOUTH AFRICA,REP OF ZR,AZQ,2,4-6
SOUTH GEORGIA IS VP8
SOUTH ORKNEY IS ..... QZ, 2, VP8
SOUTH SANDWICH IS ..... QZ, 2, VP8
S.SHETLAND IS CE9AN-AZ,QZ-Z,VP8
SOUTHWEST AFRICA(NAMIBIA) ZR3,ZS3
SOUTH YEMAN 70
SOVIET UNION:
EUROPEAN RUSSIA SOVIET FEDERATED
SOCIALIST REPUBLIC UA1,2,3,4,6
ASIATIC RUSSIAN SFSR UA9,0

ARMENIAN S.S.R UG6
AZERBAIDZHAN S.S.R UD6
ESTONIAN S.S.R UR2
FRANZ JOSEF LAND UA1
GEORGIAN S.S.R UF6
KALININGRADSK UA2
KARELO-FINNISH S.S.R UN1
KAZAKH S.S.R UL7
KIRGHIZ S.S.R ..... UMB
LATVIAN S.S.R UQ2
LITHUANIAN S.S.R UP2
MOLDAVIAN S.S.R U05
TADZHIK S.S.R UJ8
TURKMEN S.S.R UH8
UKRANIAN S.S.R UB5
UZBEK S.S.R UI8
WHITE RUSSIAN S.S.R UC2
SPAIN ..... EA
SPANISH SAHARA EA9
SRI LANKA 4S7
SUDAN ..... ST
SURINAM P2
SVALBARD IS ..... JW
SWAN IS HR0
SWAZILAND 3D6
SWEDEN ..... SJ-SM
SWITZERLAND ..... HB
SYRIA ..... YK

TAIWAN ..... BV
TANZANIA 5H3
THAILAND ..... HS
TIBET AC4
TIMOR PORTUGUESE ..... YB-YD
TOGO 5V
TOKELAU IS 2M7
TUNGA A35
TRANSKEI S8
TRANDADE & MARTIM VAZ IS PY0
TRINIDAD & TOBAGO IS 9Y4
TRISTAN DE CUNHA ZD9
TROMELIN IS FR7
TUNISIA 3V8
TURKEY ..... TA,TC
TURKS & CAICOS IS VP5
TUSCAN ARCHIPELAGO ..... IA
TUVALU IS VR8

UGANDA 5X5
UNITED ARAB EMIRATES A6X
UNITED NATIONS, GENEVA 4U
UNITED STATES OF AMERICA ..... N,W,K
URUGUAY CX

VITICAN CITY ..... HV
VENEZULA ..... YV

VIETNAM ... XV5
VIRGIN IS. ... KV4
VOLTAIC REPUBLIC ... XT2

WAKE IS. ... KW6
WALES ... GW
WALLIS & FUTUNA IS ... FW8
WILLIS IS. ... VK9Z

WINDWARD IS:
- DOMINICA ... VP2D
- ST. LUCIA ... VP2L
- ST. VINCENT & DEPENDENCIES ... VP2S

YEMEN ... 4W
YUGOSLAVIA ... YU

ZAIRE, REPUBLIC OF ... 9Q5
SAMBIA ... 9I,9J
ZANZIBAR (TANZANIA) ... 5H1

Index

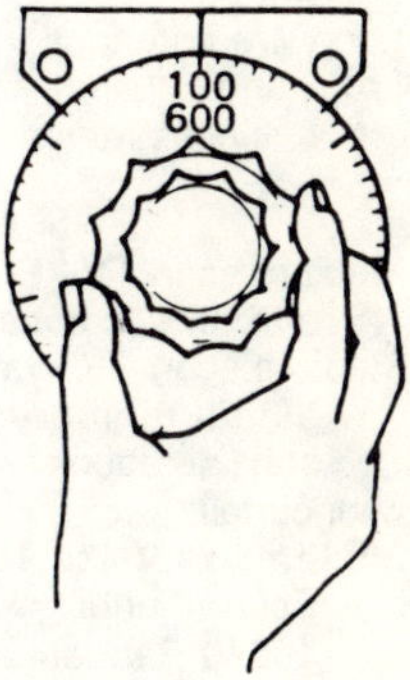